NEW POWER

NEW POWER

How it's changing the 21st century –
and why you need to know

Jeremy Heimans & Henry Timms

MACMILLAN

First published 2018 by Doubleday,
a division of Penguin Random House LLC, New York

First published in the UK 2018 by Macmillan
an imprint of Pan Macmillan
20 New Wharf Road, London N1 9RR
Associated companies throughout the world
www.panmacmillan.com

ISBN 978-1-5098-1418-3

1 3 5 7 9 8 6 4 2

A CIP catalogue record for this book is available from the British Library.

Book design by Michael Collica

Printed and bound by CPI Group (UK) Ltd, Croydon, CR0 4YY

Visit **www.panmacmillan.com** to read more about all our books
and to buy them. You will also find features, author interviews and
news of any author events, and you can sign up for e-newsletters
so that you're always first to hear about our new releases.

CONTENTS

1

WELCOME TO THE NEW
POWER WORLD

Power, as philosopher Bertrand Russell puts it, is the "ability to produce intended effects."

That ability is now in all of our hands. Today, we have the capacity to make films, friends, or money; to spread hope or spread our ideas; to build community or build up movements; to spread misinformation or propagate violence—all on a vastly greater scale and with greater potential impact than we did even a few years ago.

Yes, this is because technology has changed. But the deeper truth is that *we* are changing. Our behaviors and expectations are changing. And those who have figured out how to channel all this energy and appetite are producing Russell's "intended effects" in new and extraordinarily impactful ways.

Think of the hoodie-clad barons who sit atop online platforms a billion users strong, tweaking our daily habits, emotions, and opinions. The political neophytes who have raised passionate crowds and won stunning victories. The everyday people and organizations who are leaping ahead in this chaotic, hyperconnected world—while others fall back.

This book is about how to navigate and thrive in a world defined by the battle and balancing of two big forces. We call them old power and new power.

Old power works *like a currency*. It is held by few. Once gained, it is jealously guarded, and the powerful have a substantial store of it to spend. It is closed, inaccessible, and leader-driven. It downloads, and it captures.

New power operates differently, *like a current*. It is made by many. It is open, participatory, and peer-driven. It uploads, and it distributes. Like water or electricity, it's most forceful when it surges. The goal with new power is not to hoard it but to channel it.

To start to see how old and new power work, here are three very different stories.

#MeToo vs. Harvey Weinstein

Award seasons after award season, movie producer Harvey Weinstein ruled over Hollywood like a god.

In fact, according to *Quartz*, who analyzed Academy Award acceptance speeches between 1966 and 2016, he actually *tied* with God for the total number of times each was thanked in acceptance speeches on Oscar night—thirty four. His films garnered over three hundred Oscar nominations. The Queen made him an honorary Commander of the British Empire.

Weinstein hoarded his power and spent it like currency to maintain his vaunted position: he could make or break a star, he had huge personal capacity to green-light a project or sink it. He shaped the fortunes of an entire industry—and in turn that industry protected him even as he carried out a decades-long spree of alleged sexual harassment and assault. He controlled the media through developing a cozy mutually beneficial relationship based on the favors and access he could grant. He even won the 2017 Los Angeles Press Club "Truth Teller" award.

He buffeted himself with an army of lawyers, relying on punishing non-disclosure agreements for those who worked with him and, when necessary, paying off accusers. He hired private security firms—staffed with former spies—to dig for information on women and journalists with allegations against him. The women he preyed upon mostly kept quiet anyway, out of the very real fear of career consequences, while the men who might have stepped up stood by and did nothing, unwilling to spend their own power on a fight.

If Harvey Weinstein, and the closed and hierarchical system that held him up, tell a familiar story about old power, then Weinstein's fall, and especially what happened next, tells us a lot about how new power works, and why it matters.

In the days after news stories broke about Weinstein and his accusers, the actress Alyssa Milano shared the hashtag #MeToo to encourage women to tell their stories of sexual harassment and assault on Twitter. Terri Conn paid attention. In her twenties, as an emerging actress with a role on a soap opera, Conn had been approached by director James Toback to meet in Central Park to talk about a part. Once there, as she reported to CNN, he assaulted her.

She buried the memory for years. But with the attention on Harvey Weinstein, and the rise of the #MeToo movement, it resurfaced. She finally told her husband, and she started to act. She began by searching Twitter for women who had used both the #MeToo hashtag and #JamesToback. She found others whose stories were frighteningly close to hers. Together they formed a private Twitter group to support one another and find other survivors. Members of this group then took their stories to a journalist at the *Los Angeles Times*. Within days of an article being published, more than three hundred women came forward with stories of their own about Toback.

Conn's campaign was one of many. Almost one million tweets used the hashtag #MeToo in forty-eight hours. In just one day, twelve million Facebook comments, posts, and reactions were logged.

The #MeToo movement surged across the world like a current, with different communities adapting it to take on their own targets. In France it became #BalanceTonPorc (Denounce Your Pig), a campaign to name and shame harassers. In Italy women recounted their stories under the banner #QuellaVoltaChe (The Time That). And it moved from industry to industry. Members of Congress revealed that they, too, had been harassed by their male peers. The UK defense minister was forced to resign. The European Parliament had its #MeToo moment. Business leaders were exposed and toppled. Rallies spilled out onto the streets in cities across the world, from Paris to Vancouver. India debated an effort to expose the predatory behavior of well-known professors. An article in *China Daily* that seemed to suggest workplace harassment and assault were only Western problems was pulled after a wave of online criticism.

No one was the boss of this movement, and no one quite knew where it would go next. #MeToo had been born a decade earlier as the work of grassroots activist Tarana Burke, who encouraged women of color who had been sexually assaulted to share their experiences, peer-to-peer, with other survivors. But now the movement felt ownerless—and this was the source of its strength. Everyone from enterprising designers who created "me too" jewelry to aspiring politicians who aligned with #MeToo to seek to channel its energy.

The most striking thing about #MeToo was the sense of power it gave to its participants: many who had felt for years that they were helpless to stop longtime abusers, or had been afraid of retribution, suddenly found the courage to stand up to them. Every individual story was strengthened by the surge of the much larger current. Each individual act of bravery was, in fact, made by many.

The patient(s) vs. the doctor

The doctor looked up from his computer, stunned. "Where did you learn that word? That's my terminology. When did you go to medical school? I can't see you as a patient anymore if you're going to go on the internet and just learn stuff that you shouldn't be learning."

Then the doctor fired his patient.

The offensive word was "tonic-clonic." His patient had let him know that she thought she had experienced a secondarily generalized tonic-clonic seizure. (In the past, she and her doctor had referred to these moments as "space-outs," regular seizures that had been causing her serious concern.)

This patient had learned about her condition through Patients-LikeMe, an online community of over 500,000 people living with more than 2,700 diseases, each of whom shares their personal medical data and experiences with others on the platform, creating tens of millions of data points. Think of it as a massive support group, learning community, and data set, all rolled into one. Patients on the platform have even worked together to crowd-source their own drug trials, such as when a group of ALS patients conducted a test of lithium as a treatment in a fraction of the time it would have taken the health authorities.

Letitia Browne-James, another member of the community, stumbled upon PatientsLikeMe "out of desperation." She had suffered from epilepsy her whole life, enduring frequent and debilitating seizures that were just getting worse. She feared having a seizure in school or in church, while she was acting or dancing, or, as she got older, on a date.

After she met her future husband, Jonah James Jr., she worried about her wedding day. "I prayed really hard, just asking God to allow me to let me make it through that day without having a seizure," she said.

While her neurologist kept on prescribing the same old medications, she began to confer with community members on the platform, learning for herself about why certain drugs weren't working, and trying to figure out what other options might be possible. Chasing any kind of hope, she was told of the promise of brain surgery as a treatment for people with epilepsy. She discovered that 83 percent of her fellow patients on the platform had reported positive outcomes from this type of treatment, yet it was something she and her doctor had never even discussed.

So this patient fired *her* doctor. As a parting request she asked for the name of an epileptologist—the type of specialist she had learned about from her patient community. The doctor flipped through papers on his desk and gave her a name. She was aghast. "He had had that information there all the time," she said.

Letitia underwent the surgery. She has now had more than five years without a single seizure. And she has mentored many others on PatientsLikeMe, helping them take control of their health.

The doctors in these stories live in a world that runs on old power. They have trained rigorously to develop their expertise. And for good reason: they are dealing with matters of life and death. But in doing so, they have become accustomed to being the keepers of medical knowledge, distanced from their patients by a hyphenated lexicon and inscrutable prescriptions. The patients have discovered new power. They act to improve their own conditions, surrounded— and rallied—by a crowd of like-minded people. They try things out, swap journal articles, and track each other's progress. They share their data, ideas, and compassion. Their worlds have opened up— and no doctor can put that genie back in its bottle.

The schoolgirl vs. the State Department

Aqsa Mahmood grew up part of a moderate Muslim family in Scotland. She attended good private schools and loved Harry Potter. She was described as someone who didn't know which bus to take to find her way to downtown Glasgow.

Yet, over time, she became a "bedroom radical," falling into a dark online ecosystem of persuasive content and seductive recruiters. Then one day in November, when she was just nineteen years old, she disappeared. When her parents next heard from her, four days later, she was calling them from the Syrian border.

But this was not the end of her story. Having been recruited into ISIS, she now turned recruiter, mastering the tools of online engagement and enticing others to follow her example. She built a close-knit girl-to-girl network, sending encouragement and offering practical advice for wannabe jihadi women who were preparing to make the journey to Syria: "If I could advise you to bring one thing it would be organic coconut oil (maybe grab an extra jar for me as well lol). This is such a helpful product with multi-use—

body moisturiser/hair oil, etc." When three normal and well-liked girls from Bethnal Green, London, plotted their own departure for Syria, it was Aqsa Mahmood to whom they reached out on Twitter.

While Aqsa used intimate, peer-to-peer methods to win over recruits, the U.S. government took a very different approach to try to dissuade them. It printed thousands of cartoons of ISIS recruits being fed into a meat grinder and dropped them out of an F-16 fighter jet as it flew over ISIS strongholds in Syria (an approach that had first been widely used a hundred years earlier, during World War I). It tried a digital approach, too, in an attempt to match the Islamic State's online savvy, creating a rather bossy Twitter account—replete with an ominous State Department seal—that instructed potential jihadis to "Think Again Turn Away!" This was perhaps not the most persuasive messenger if you're trying to pull radicalized people back from the brink.

Here again we see old power meeting new power. The U.S. government was relying on a trusty old power playbook, using its superior position to literally drop ideas from on high. Even when using social media, its default is not to engage, but to command. Aqsa is doing something very different. Her makeshift, metastasizing network is participatory and peer-driven. It moves not top-down, but *sideways* from girl to girl. It is new power at its most effective, and most terrifying.

THE INGREDIENTS OF NEW POWER

What the #MeToo movement, our patients, and a Scottish schoolgirl all have in common is that they figured out how to use today's tools to channel an increasing thirst to participate.

People have always wanted to take part in the world. Throughout history, movements have surged, people have organized collectively, communities have built collaborative structures to create culture and conduct commerce. There has always been a dialectic between bottom-up and top-down, between hierarchies and networks.

But until recently, our everyday opportunities to participate and agitate were much more constrained. Thanks to today's ubiquitous connectivity, we can come together and organize ourselves in ways that are geographically boundless and highly distributed and with unprecedented velocity and reach. This hyperconnectedness has given birth to new models and mindsets that are shaping our age, as we'll see in the pages ahead. That's the "new" in new power.

A popular thread on Reddit, the link-sharing platform, crowd-sourced memories of growing up in the 1990s, when life felt very different. For those who were there, the posts offered warm nostalgia. For those who weren't yet born, it told stories of an alien world: The anxiety of waiting for your yearbook photo to arrive, which was "the only time you saw a picture of you and your friends at school." You only got one shot to get that right, and you never knew how it would turn out. The tension of calling the local radio station, requesting your favorite song, and then waiting, fingers poised on the record button of your tape cassette player, to capture it when it came on. The excitement of stopping by the Blockbuster Video store to rent a movie on the way home. The frustration of going to the library and finding the one book you need has already been taken out or "should be in the stacks but can't be found." The tedium of doing math without a calculator because they were banned, the sturdy reasoning being "you won't have a calculator in your pocket all the time when you grow up."

Of course, we now have much more than a calculator in our pocket. In today's world, we all have our hands (quite literally) on what we can think of as a new *means of participation*. And this isn't just changing what we can do, but how we expect to engage.

These new means of participation—and the heightened sense of agency that has come with them—are a key ingredient in some of the most impactful models of our time: big businesses like Airbnb and Uber, China's WeChat or Facebook; protest movements like Black Lives Matter, open software systems like GitHub; and terrorist networks like ISIS. They are all channeling new power.

Think of these as *new power models*. New power models are enabled by the activity of the crowd—without whom these models are just empty vessels. In contrast, old power models are enabled by what people or organizations own, know, or control that nobody else does—once old power models lose that, they lose their advantage. Old power models ask of us only that we comply (pay your taxes, do your homework) or consume. New power models demand and allow for more: that we share ideas, create new content (as on YouTube) or assets (as on Etsy), even shape a community (think of the sprawling digital movements resisting the Trump presidency).

To grasp the essential difference between old and new power models, think of the difference between the two biggest computer games of all time, Tetris and Minecraft.

You will likely remember the block-based game Tetris, which exploded with the Gameboy craze of the 1990s. The way it worked was simple. Blocks fell down from the top of the screen and the player's job was to make them fit into neat regular lines. They came down faster and faster until the player was eventually overwhelmed. In old power fashion, the player had a limited role, and you could never beat the system.

New power models work more like Minecraft, now the *second* biggest game of all time. Like Tetris, it is a clunky block-based game. But it operates very differently. Instead of a model built on top-down compliance, it is a game built from the bottom-up, with players around the world co-creating worlds together, block by block. It relies entirely on participatory energy. In the world of Minecraft, you will find houses, temples, and Walmarts; dragons, caves, boats, farms, and roller coasters; working computers made by engineers; forest fires, dungeons, cinemas, chickens, and stadiums. The players set their rules and create their own tasks. There is no "manual"; players learn from the example—and often the home-made videos—of others. Some players (known as "modders") are even entrusted with the capacity to alter the game itself. Without the actions of the players, Minecraft is a wasteland. A key dynamic in the world today is the mutual incomprehension between those raised in the Tetris tradition and those with a Minecraft mindset.

THE MISSION OF THIS BOOK

The future will be a battle over mobilization. The everyday people, leaders, and organizations who flourish will be those best able to channel the participatory energy of those around them—for the good, for the bad, and for the trivial.

This matters in the daily lives of all of us

Since we first wrote about these ideas in the *Harvard Business Review,* it has been inspiring to see people in so many different sectors using them to reimagine their worlds, from librarians to diplomats to health workers. In the chapters ahead, we will tell stories of organizations and individuals who understand these new dynamics. We'll unpack how the Lego company saved its brand by turning to the crowd. We'll consider how TED grew from an exclusive conference into one of the biggest ideas communities in the world. We'll look at how Pope Francis is trying to shift the nature of his church by empowering his flock.

We'll introduce some lesser-known examples, too: nurses banding together to cut down on bureaucracy and improve patients' lives (and their own job satisfaction); a car company that turns to its customers to design its vehicles; a successful media company built, funded, and shaped by its readers.

Whether you are a historian yearning to share your knowledge in a post-truth world, a determined parent running for your local school board, or a creator wanting to get a new product off the ground, there are a range of distinctive new capabilities that people and businesses need to discover.

The skills in question are often misunderstood as the ability to self-promote on Facebook or as *Snapchat for Dummies.* But new power is about much more than just new tools and technologies. As the State Department showed us in their failed online sparring with the Islamic State, many are still deploying these new means of participation in profoundly old power ways. This book is about a different approach to the exercise of power, and a different mind-set, which can be deployed even as particular tools and platforms go in and out of fashion. How do you create ideas that the crowd grabs on to, makes stronger, and helps spread? How do you operate effectively within an organization in which your (perhaps younger) peers have internalized new power values like radical transparency or constant feedback? How do you create an institution that inspires an enduring, mass following in an era of much looser, more transitory affiliation? How do you switch between old and new power? When should you blend them together? And when will old power actually produce better outcomes?

This book will answer these questions—and more—drawing on examples from some of the most inspiring new power success stories (and some of the big cautionary tales) from around the world.

This matters for society at large

New power is here to stay and is, in many sectors, ascendant. In the right hands, it is doing wonders: the crowd-sourced drug trials; the fast-growing movements in the name of love and compassion. Yet in the wrong hands, as we see with ISIS or the growing hordes of white supremacists, these same skills can be enormously destructive. The tools that bring us closer together can also drive us further apart.

Those who are building and stewarding vast platforms that run on new power have become our new elites. These leaders often use the language of the crowd—"sharing," "open," "connected"—but their actions can tell a different story. Think of Facebook, the new power platform that most of us know best. For all those likes and smiley faces we create using what the company calls our "power to share," the two billion users of Facebook get no share of the vast economic value created by the platform. Nor any say in how it is governed. And not a peek into the algorithm that has been proven to shape our moods, our self-esteem, and even some elections. Far from the organic free-roaming paradise the early internet pioneers imagined, there is a growing sense that we are living in a world of *participation farms,* where a small number of big platforms have fenced, and harvest for their own gain, the daily activities of billions.

The stakes are high for democracy as well. Many hoped that surges of social media alone would topple dictators. But in fact a new kind of strongman is on the rise in many parts of the world, supercharged by the very tools some believed could only democratize. Take Donald Trump. Trump became the leader of a vast, decentralized social media army who took cues from him—and who in turn fed Trump new narratives and lines of attack. It was a deeply symbiotic relationship. He retweeted his most extreme supporters. He offered to pay the legal fees of supporters who punched protesters at his rallies. He drove the intensity of his crowd not by insisting they read his talking points, but by empowering them to activate around his values. Think of him as a *Platform Strongman,* mastering new power techniques to achieve authoritarian ends.

In the chapters ahead, we will explain the dynamics that make participation farms and platform strongmen possible. Critically, we'll also showcase stories of their antidotes: those new models that genuinely shift and distribute power to more people, including the least powerful among us. We'll meet pioneers who are imagining ways to reinvent democracy, not undermine it, finding ways to transform citizens from hostile outsiders to co-owners and valuable players in the work of government. We'll also visit traditional institutions in vital parts of society that are taking the tough turn from old to new power. We hope this book equips those fighting for a more open, democratic, and pluralistic world with the tools they need to prevail.

This book is grounded in our own experience creating new power models and trying to bring more participation to more people. Henry launched #GivingTuesday, a philanthropic meme that become a movement, raising hundreds of millions of dollars for charities around the world. Jeremy created a technology-powered political movement in his home country of Australia as a twenty-something that became the biggest in the nation, and he has since helped to launch many more movements around the world via his organization Purpose, headquartered in New York. We've seen the potential and pitfalls of new power up close, and now we want to share what we've learned. We've been working together, and engaging with businesses and communities, to dig deeper into what's changing, why, and what we can all do about it.

In the pages ahead, we'll share what we've discovered.

THINKING OLD POWER, THINKING NEW POWER

"THE LAB IS MY WORLD" VS. "THE WORLD IS MY LAB"

NASA's Johnson Space Center is legendary for taking on great challenges. When the Apollo astronauts broadcast "Houston, we've had a problem," it was the Johnson Space Center they were calling.

In 2010, though, NASA faced the danger of having its budget slashed by Congress, its utility implicitly questioned. It was under scrutiny, too, for a lack of imagination. As the head scientist at Johnson told his troops, "HQ are telling us we are not innovative enough, we need to show they're wrong."

So they began to experiment with something called "open innovation." Open innovation is the concept of enlisting the crowd to help solve your problem. Unlike the old power method, where a small number of experts have exclusive access to tools, data, and machines, the goal of open innovation is to invite everyone to engage. The story of NASA's efforts to open up comes from the research and insights of NYU professor Hila Lifshitz-Assaf. She spent three years deep inside the agency; a fly on the wall for a period of dramatic—and tense—change.

The Johnson Space Center's effort was led by its Space Life Sciences Directorate (SLSD), under doctor and flight surgeon Jeffrey Davis. The directorate picked out fourteen strategic research and development challenges and laid them out on open innovation platforms for the world to take on. Three thousand people in eighty countries responded, ranging from recognized experts to unrecognized weekend enthusiasts.

The initial results were impressive. The typical traditional R&D (research and development) cycle was three to five years, yet the crowd was turning problems around in three to six months. And not only did the crowd produce quicker solutions, at much lower cost, the quality of its work was significantly higher than expected.

One solution stood out above the others and became a symbol of the promise of the approach. It addressed a serious problem in heliophysics: how hard it was to forecast solar storms effectively. Solar storms are huge bursts of energy from the sun that travel toward the earth at three million miles per hour. Obviously, the ability to avoid them is a high priority for those traveling the solar system. Yet for all of the work done by experts around the world—at NASA and elsewhere—the best models allowed for forecasts at one to two hours' notice and with a 50 percent level of accuracy.

And now Bruce Cragin, a semi-retired telecommunications engineer from New Hampshire, who wasn't even a heliophysicist and didn't have access to anything like the tools at NASA, had submitted an algorithm that allowed for prediction *eight* hours in advance, with a *75 percent* level of accuracy.

This breakthrough kicked off a huge wave of enthusiasm among NASA's senior leadership, grabbing the attention of the national media and even piquing the interest of the White House. So NASA doubled down.

Jeffrey Davis pulled together a much larger group and rolled out a special workshop to unite the team around the promise of this new approach. The day kicked off with a lot of excitement and a bold claim from one of the workshop leaders: "This puts you on the edge of many organizations—in many ways of all organizations in the world—trying to figure out this open innovation perspective."

Yet rather than cueing a wave of enthusiasm, the meeting unfolded into mayhem. Lifshitz-Assaf describes the scene at the workshop: "The tensions, debates and forces unleashed on that day, led to a very different trajectory than that planned. The intensity of fears and resistance expressed in the room throughout the day was out of the ordinary." The rough translation of this academic analysis is "people went nuts."

What had happened? Why had a promising opportunity led to such intense feeling and division? Over the months ahead, Davis and his team, bruised but determined, pushed on with their open innovation efforts. And they began to see two very different factions form.

One camp viewed it all as a waste of time, a nuisance and a threat. They grumbled about the budgetary impact of the new work. They nitpicked technical details. Some refused to discuss problems they were struggling with, "fearing it might be transformed into an Open Innovation Challenge." Some turned saboteur, dissuading their colleagues from taking part. Others showed enthusiasm in public, but provided the scantest details for the crowd to engage with and ignored their ideas when they arrived. One team even went into full denial mode, "withholding the fact it was actually participating in open innovation."

The other camp saw opportunity. They created new processes and approaches to get the best out of their crowd. They invented tools that opened up their labs for knowledge to flow in *and* out. Some left their roles altogether to set up "open NASA," a new unit dedicated to fostering the work. One team founded the now annual "Space Apps hackathon"—perhaps the most global of all such efforts—which in 2017 brought together 25,000 people in 187 locations in 69 countries over three days to engage with some of space's trickiest challenges. Another built an open platform to engage with a community of space experts and citizen scientists worldwide. Today, this appetite for open innovation at NASA has grown so much that there is now a senior policy adviser in the Office of the Chief Scientist with responsibility for engaging citizens to add value across the whole agency.

Were this any normal legacy institution, you might assume that this divide had something to do with people fearing new technology. But this was clearly not the case. After all, these were literally rocket scientists. Nor was the division about age. Or experience. Or reputation. The populations of the two camps looked pretty similar.

What *was* behind this big divide were two very different ways of thinking.

The first group had what we call *old power values*. They came from a world with clear boundaries between "us" and "them," where only the lab-coated and credentialed were equipped to solve the mysteries of the cosmos. As one leading scientist explained, the resistance to open innovation "is really intrinsic, the history of the scientific method goes against it . . . In our training, trying to solve problems in the scientific method was: I take in all this information, I synthesize it, I do analysis and I come to some conclusion and so to reach out to other people to solve it, it's like cheating!"

This group believed deeply in the value of expertise. Their own identities grew out of a tradition that venerated individual moments of genius. Archimedes leaping from his bath. Newton getting an apple on his noggin. Their instinct was to hoard information about their work, not expose it to the scrutiny of an unqualified crowd who might not play by the established rules of scientific inquiry and debate. They had some reason to be skeptical: many experiments in open innovation and crowdsourcing end up failing. These were people who had in many cases made enduring, decades-long commitments to NASA—and they weren't going to let some dilettante swan in and supplant them. Professional privileges and knowledge were hard-won currency. You are what you have amassed.

Tellingly, Lifshitz-Assaf notes that when people in this group were asked about open innovation, they would often—without prompting—start talking about "why they joined NASA, who they are and how they were trained." They'd start talking about their PhD mentors, pulling out their research papers and showing her the various professional artifacts from their careers, a demonstration of the sweat and hard work of many years. What she found odd about this, of course, was "I didn't ask them about *them*, I asked them about Open Innovation."

The looming crowd, she concluded, presented a threat to their core identity. This was a group for whom the answer to "Houston, we've had a problem" could never be "Stand by, Apollo, we're going to crowd-source that and see if any semi-retired telecommunications engineers in New Hampshire have any insights."

The second group had *new power values*. They were more open to collaboration, believed in the possibility of crowd wisdom, and wanted to open up their world to let others join in. They decided that their teams would be stronger if they could find ways to create discrete assignments that anyone around the world could help out with. For this group, even the watercooler tales they told began to change. One popular story emerged of an engineer who wanted to identify a breakthrough medical device that could be used on the international space station and found it by reaching out on YouTube. These scientists stopped thinking "The lab is my world" and started thinking "The world is my lab."

A TALE OF TWO MINDSETS

Old power and new power values are not clashing just at NASA. More broadly, two very different mindsets are doing battle in today's world.

Old Power Values	New Power Values
Formal (representative) governance, managerialism, institutionalism	Informal (networked) governance, opt-in decision-making, self-organization
Competition, exclusivity, resource consolidation	Collaboration, crowd wisdom, sharing, open-sourcing
Confidentiality, discretion, separation between private and public spheres	Radical transparency
Expertise, professionalism, specialization	Maker culture, "do-it ourselves" ethic
Long-term affiliation and loyalty, less overall participation	Short-term conditional affiliation, more overall participation

The twentieth century was built from the top down. Society was imagined as a great machine, intricately powered by big bureaucracies and great corporations. To keep the machine humming, ordinary people had critical, but small and standardized, roles to play. Do your drills. Say your prayers. Learn your multiplication tables. Serve your time. Sit for your yearbook photo. Many of us were relatively content to play a minor role in a larger process. Yet the rise of new power is shifting people's norms and beliefs about how the world should work and where they should fit in. The more we engage with new power models, the more these norms are shifting. Indeed, what is emerging—most visibly among people under thirty (now more than half the world's population)—is a new expectation: an inalienable right to participate.

The YouTube creator with her own robust set of followers approaches the world expecting to be a creator more than a consumer. A person who stitches together a living as a "producer" for Task-Rabbit, Lyft, or other on-demand services may become less reliant on, and more skeptical of, traditional economic intermediaries. An employee who gets the thrill of unlimited creativity and immediate validation from her online community can find her mundane daily workplace projects, which her boss rarely comments on, especially unfulfilling. A citizen who becomes deeply involved in a crowd-funded project she has backed in her neighborhood may become disillusioned or disengaged when her interactions with her local government are mainly in the form of summonses and paperwork. Participatory experiences are multiplying across all of our lives and further shaping this new power mindset.

It is important not to see all this in normative terms. This is not a case of "new power values = good," "old power values = bad." After all, there are many times we might choose old power values over new. If you are having a root canal done, for example, you'd almost certainly value the expertise of an endodontist with two degrees and fifteen years' experience, over a group of collaborative amateurs and wannabe weekend "makers" who had crowd-sourced their drills and picked up the technique from an anonymous post on Reddit. And while these two mindsets often come into conflict, we shouldn't see the values as a binary. It's better to see them as a spectrum and to consider where your beliefs, and those of the organizations you are part of, might sit along them. Let's unpack each set of values.

Formal vs. informal governance

"We've heard people say that teachers have no business going rogue and trying to select their own books, technology, and classes—and citizens have no business deciding what is worthy. And yes, we have a position on that, and a response to people who raise that question. Screw you. We believe in teachers."

These are the words of Charles Best, the founder of DonorsChoose, which allows teachers to take matters into their own hands, and raise funds for what *they* see as their classrooms' needs—from supplies to laptops. The site has seen over two million "citizen donors" step up to give more than $400 million to help 18 million public school students, whose needs are often left unmet by under-resourced U.S. public schools. It's one of the earliest and most successful crowdfunding platforms. Classes of kids will often send adorable thank-you pictures and notes to their funders, one of many ways the platform has become so compelling.

Yet not everyone is enthusiastic about Best's approach. In a profile of DonorsChoose, *Fast Company* quotes Columbia University professor of political science Jeffrey Henig expressing an old power view of how education funding decisions should be made: "We have vested school boards, superintendents or mayors' offices with authority to make decisions about schooling because we understand they will be made out in the open, where questions of conflicting values are negotiated and compromises are made," he says. "It will be a collective process embedded with democratic procedures and discussion."

Henig is articulating the virtues of more centralized, formal, and representative governance, over the vicissitudes of the crowd. He thinks education, as a public good, should be subject to resource-allocation decisions that consider what's best and fairest for the system as a whole, rather than rewarding the most energetic or persuasive teacher. But Best is backing new power values, believing in the just-get-it-done, devolutionary impulses of crowdfunding as a way to help kids now. It's Henig who is defending formal governance processes and the idea of delegating authority to elected representatives, while Best champions the virtues of direct participation and individual agency. (Notice that in crowdfunding models, more participation does not necessarily lead to more equal representation or inclusion; sometimes new power can mean less of both.)

Those with a new power mindset have an aversion, which often comes with a dollop of disdain, for the centralized bureaucratic machines that drove the old power world. They prefer more informal, networked, and opt-in means of getting things done. They despair of those who take their dusty places at the biweekly meeting of the standing deliberative committee on multi-sector-high-level-decision-making. More flash mob than United Nations, this philosophy stands in contrast to the twentieth century belief in managerialism and institutionalism as the way to get things done.

At the extreme, the belief in "informal" governance manifests in those Silicon Valley–led dreams of floating island paradises, "an opt-in society, outside the US, run by technology." This is the kind of place, as one prominent Valley entrepreneur has advocated, where a "Yelp for Drugs" replaces the U.S. Food and Drug Administration, with doctor ratings and patient testimonials supplanting regulations and formal protections.

Competition vs. collaboration

New power models, at their best, reinforce the human instinct to cooperate (rather than compete) by rewarding those who share their own assets or ideas, spread those of others, or build on existing ideas to make them better. Many new power models such as Airbnb are driven by the accumulated verdict of the community. They rely on reputation systems that ensure that, say, rude or messy guests on the platform have trouble finding their next places to stay. A commonsense strategy to build followers on Twitter is to retweet and promote the thinking of others, in the expectation that they will reciprocate. In a networked world, collaborating with your neighbor or someone on the other side of the world is both much easier and more frequently rewarded. The most successful open-source software engineers are the ones who collaborate best; they build on and improve the work of their peers, even when there's no obvious immediate benefit to them. Even big companies like GE are now talking up "a radical shift in everyday working behavior" toward collaboration.

In contrast, those with old power values celebrate the virtues of being a great (and sometimes ruthless) competitor, defined by your victories. Dividing the world into winners and losers, this mind-set considers success a zero-sum equation. It is the classic thinking behind much of corporate life and essential to the culture of sales teams in almost every industry. Donald Trump is steeped in these values, as is Uber, especially under the leadership of its co-founder and former CEO Travis Kalanick. Despite its new power model, Uber has a track record of sabotaging its competitors, intimidating journalists, and hoodwinking government regulators to come out on top. In a leaked document that detailed what it looked for in employees, Uber highlights "fierceness" and "super-pumpedness," all part of a "hustle" culture.

It's worth noting that while norms around collaboration and "sharing" are now all the rage in our business and culture, that doesn't mean they always produce better outcomes. A recent study in *Applied Psychology* found that "cooperative contexts proved socially disadvantageous for high performers"—who find themselves ostracized by the rest of the group.

Confidentiality vs. radical transparency

In the leaked transcripts of presidential candidate Hillary Clinton's paid speeches after leaving the State Department, she perfectly articulates old power norms about how information should flow: "I mean, politics is like sausage being made. It is unsavory, and it always has been that way, but we usually end up where we need to be. But if everybody's watching, you know, all of the back room discussions and the deals, you know, then people get a little nervous, to say the least. So, you need both a public and a private position . . ."

To many young people, this pragmatic rationale for lack of transparency or straightforwardness just isn't acceptable anymore. In an era in which young people are sharing the most intimate details of their lives on social media, it shouldn't be surprising that in the workplace they are now demanding that their bosses share information previously considered strictly confidential, like company-wide salary information. The separation between public and private spheres, so prized in an old power world, is dissolving—and is being replaced by an ethos of radical transparency. (The great irony, of course, is that the more transparent people are in how they lead their lives, the easier it is for unseen forces to track their actions and shape their behaviors.)

The big clash here is between the "need to know" mindset, which instinctively keeps information away from the public for its own protection, and a rising "right to know" expectation, where new power thinkers demand openness from institutions as a default. In the first case, experts and authority figures decide what filters information deserves; in the second the filters don't exist.

Across sectors, the old power world is absorbing a sustained attack on the hidden, with its dirty laundry increasingly exposed—by WikiLeaks and in the Paradise Papers, for example—and its daily performance monitored: the once-untouchable professor now must tolerate her teaching being rated online at any hour by neophyte students.

In an age when it is ever harder to guard secrets and avoid scrutiny, some leaders and institutions are embracing radical transparency, if only as a preemptive strategy. To borrow a phrase from the world of protest, they are choosing to *occupy themselves* before they themselves are occupied. Our favorite example of this approach is that of 2018 Arizona gubernatorial candidate Noah Dyer, whose campaign website includes a section called "Scandal and Controversy." In it, he declares the following: "Noah has had both deep and casual sexual experiences with all kinds of women . . . He's had group sex and sex with married women. He has sent and received intimate texts and pictures, and occasionally recorded video during sex. Noah has always been forthright with his partners. All of his relationships have been legal and consensual . . . Noah is unapologetic about his sexual choices, and wishes others the same safety and confidence as they express themselves." We won't hold our breath for a similar declaration from Mitch McConnell.

Experts vs. Makers

"It really comes down to just being able to make something. Make something that's important . . . Always be creating . . . For me it's that sweet, sweet beautiful toasted bread. I'm Andy Corbett, and I make toast."

That's a snippet from a 2016 ad that made light of "maker culture." Glorifying being a "creator" rather than just a passive consumer, and best of all a hybrid—a coder-designer-engineer-doctor-musician-vet—is a big theme in new power culture.

We can think of maker culture as the "do-it-ourselves" mindset (a phrase of Scott Heiferman, the CEO of Meetup, which reflects the idea that we should be makers, but in collaboration with others). We see this come alive in everything from "amateur" pornography, to people printing shoes in their garages using online templates, to "GynePunks," groups of women taking care of their own reproductive health via homemade incubators and 3D printable speculums, to the surge in fan fiction communities like Wattpad. Makers are less dependent on institutions. They figure out how to avoid the intermediaries.

This trend is allied with a shifting view of expertise. Restaurateurs, filmmakers, hoteliers, artists, and writers used to live in fear of the all-powerful critic whose expert opinion could tank or make their enterprise. Today, these elites still retain great influence, but increasingly we look to one another for direction. Our world is a little more Yelp and a little less Frommer's. In fact, over the last decade, the Edelman Trust Barometer has measured significantly increasing trust in "people like me," which has reached higher levels than public trust in academic experts or doctors.

This clash was dramatized by the run-up to the Brexit vote, when the "we know what's good for you" IN campaign, seen as led by economic and cultural elites and technocratic "experts," was swept away by the populist OUT vote. Michael Gove, a senior government minister who favored Brexit, exploited this sentiment, declaring during the campaign that "people in this country have had enough of experts," pitting the economists who predicted Brexit would be bad for Britain against everyday people. This ploy enraged experts like the particle physicist Brian Cox, who said in response: "It's the road back to the cave. Being an expert does not mean that you are someone with a vested interest in something; it means you spend your life studying something. You're not necessarily right—but you're more likely to be right than someone who's not spent their life studying it."

Long-term vs. transient affiliation

Robert Putnam's now-classic book *Bowling Alone* heralded declining civic health in the United States. To track this, he used a range of measures including attendance at a public meeting on town or school affairs, service on a committee for a local organization, and membership in a chapter-based organization. These standards speak to old power values around affiliation, which associate "taking part" with regular attendance and adherence to organizational charters and party platforms.

But just as we cannot fairly measure the vitality of the media industry by the number of current cardholders at Blockbuster Video, we cannot fairly measure the health of civil society today by the number of members of clubs. Those with new power values are less committed but more affiliative—and that's a paradox many old power institutions are now grappling with.

Since the advent of the internet we've seen a huge new wave of joining, affiliation, and participation, though not on the terms Putnam was seeking. New power loves to affiliate, but affiliation in this new mindset is much less enduring. People are less likely to be card-carrying members of organizations or to forge decades-long relationships with institutions, but they are more likely to float between Meetup groups or use social media to very visibly affiliate with a range of causes, brands, and organizations, and rally their friends to do the same. They tend to opt in at particular moments, and then opt out again. We shouldn't confuse this with a lack of engagement. Rather it is a different way of taking part. This shift has big implications for organizations large and small.

NEW POWER VALUES IN ACTION—WELCOME TO THE HOUSE FOR ALL SINNERS AND SAINTS

At the House for All Sinners and Saints in Denver, Colorado, we see many of these new power values in action.

What makes pastor Nadia Bolz-Weber different isn't the tattoos of religious stories all over her body, nor is it her media-friendly rags-to-dog-collar story. It's her congregation's simple philosophy: "We're anti-excellence, pro-participation."

Her church, the House for All Sinners and Saints, is booming, packed with millennials, clearly speaking to a demographic that mainstream Christianity often struggles to reach. Much of that success comes from the way she has structured her church to accommodate and celebrate new power values.

If you stop by one of its services, even as a first-time church-goer, you might find yourself delivering the liturgy. Each service is performed by between fifteen and eighteen ordinary attendees who grab a part as they arrive and then collaborate to lead the congregation. And why is this so important? As Bolz-Weber explained to us, it sends a critical message. "It's like telling people that we trust you with the holy things right away, just because you showed up."

Consider the way they plan their Ash Wednesday and Lent services—some of the most important in the Christian calendar.

The mainstream approach to this, as Bolz-Weber tells us, would be to create a formal governance structure: "I need eight people to be the Liturgy Committee, the Worship Committee, it's the second Tuesday of the month for an hour and a half and it's a two-year commitment." Yet she knows how many people this would appeal to in her congregation: none.

Instead, they just throw out some times and see who turns up. Then they let those who arrive get on with it. Last year, three of the people who planned the Ash Wednesday service had never actually attended one before. Their first service was the one they had co-created.

This belief in collaboration between everyday churchgoers is at the heart of the church's success. As Bolz-Weber puts it, "We don't really care about doing things well, we just care about doing them together." This approach extends to the somewhat chaotic singing, too. There is no choir for services; all singing is taken on by those who attend. They even pray in a more collaborative way, using Google groups to allow people to offer their thoughts for those in need whenever the urge takes them—on the bus, after work, during work, whatever suits.

"House," as the church is lovingly referred to, couldn't be more different than most mainstream churches. (Bolz-Weber wryly describes these mainstream congregations as being made up of "basically twelve old people and their parents.") What House does so well is to offer increased agency, flattened hierarchy, and a joyful embrace of diversity, the opposite of the quiet-in-the-pews, top-heavy, single-note experience that many find when they go to a traditional place of worship.

Some will roll their eyes at House, but Bolz-Weber and her peers are filling what is a very big gap. As a Pew survey reports, "Among the youngest Millennials . . . fully one-third (34%) are religiously unaffiliated, compared with about one-in-ten members of the Silent Generation (9%) and one-in-twenty members of the World War II–era Greatest Generation (5%)." Many of these young people say they believe in God: they just aren't finding Her in the rituals of old power institutions.

House shows us new power values fully operationalized. And whether you identify with these values or they make you want to tear your hair out, we all need to understand this new power mindset. In the schools, workplaces, hospitals, and even battlefields of the twenty-first century, more and more people are adopting it. And as they do, their expectations about their own participation and agency keep getting higher.

THE NEW POWER COMPASS

Those businesses and organizations that rely on new power *models* are not necessarily embracing the new power *values* we have unpacked in this chapter. In fact, we see different combinations of new and old power models and values that reflect very different strategies for survival and success. We can more clearly understand a lot of the successful (and some of the more challenged) organizations of our era by using the new power compass.

The New Power Compass

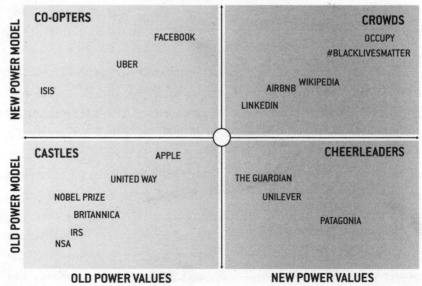

Open

The horizontal axis tracks the values of an organization: whether it exhibits new or old power values. The vertical looks at its model: whether it is a new power model designed and structured to encourage mass participation and peer coordination or an old power model that asks us to do little more than comply or consume.

Black Lives Matter, for instance, has a new power *model*: it is highly decentralized, lacking an organizational owner or traditional leaders. It has inspired coordination and participation among people all over the United States. It also displays new power *values*: it gives huge leeway to its supporters to adapt its message and has been highly transparent about its decision-making, such as when a loose-knit collection of groups came together to produce its first policy platform. For all these reasons, it fits squarely in the **Crowds** quadrant.

We can place Airbnb in the same quadrant, though much closer to the center than Black Lives Matter. It has an amazingly effective new power model, with hosts in 65,000 cities and more than three million listings. It tends toward new power values, too, championing the individuality of its hosts, encouraging collaboration and self-organization in its community, and promoting transparency and trust as core community norms. As Airbnb continues to grow and faces regulatory challenges and pressure from investors, a key question is whether it will stick to those values or gravitate to old power.

There are strategic dangers in this quadrant. Think about Occupy, the protest movement against inequality that erupted in 2011 and spread around the world, which was a study in both new power values and new power model: distributed, decentralized, collaborative, and radically open. Yet its commitment to achieving consensus at all costs, and its opposition to any kind of institutionalization, made it excruciatingly difficult to get decisions made or to advance a more concrete program of change.

Beneath the Crowds quadrant, we find the **Cheerleaders**. These are organizations with old power models that embrace new power values. Take Patagonia, whose model is quintessentially old

production　open

power—it makes stuff like great jackets, with no involvement from outside its four walls—but which has embraced a highly collaborative relationship with its consumers in other ways, enlisting them to champion causes like climate change and even to attack consumerism itself. It has shown a strong commitment to transparency, opening up information about its supply chain, shining a light on topics that many big companies would prefer to keep under wraps, like the wages of its garment workers and the environmental impact of textile production.

Another Cheerleader is *The Guardian,* a venerable old media brand, which is trying to reimagine its work while resuscitating its finances. *The Guardian* has championed the notion of radical transparency by publishing state secrets, and has come under heavy criticism from governments, and even some other media companies, for doing so. It has also been transparent about its own problems— notably its falling ad revenue—and has been relentless in asking its readers to help it by becoming "members" and making a monthly donation. Impressively, in 2017 it claimed to have 230,000 members; a higher number than it has subscribers. It now pulls in about the same amount of money from its readers and members as it does from its advertisers. It has been experimenting with new power models by asking its readers to directly participate in its work, such as its efforts to lobby Bill Gates to divest from fossil fuels, and its database, "The Counted," which is compiled by readers and journalists to document all the people killed by law enforcement in the United States. But it counts as a Cheerleader rather than a Crowd because, despite its embrace of new power values, it remains mostly a traditional media model based on "download."

The most familiar—and populated—of the quadrants is the **Castles**, those organizations with old power models and values. We all know them well—many of us work for them, from assembly line manufacturing to advertising. A classic example might be the National Security Agency, which lives in the shadows and whose main interest in mass participation lies in secretly surveilling masses of people. Or consider the Nobel Prize committees, small groups of experts who get together once a year behind closed doors to decide who the smartest people in the world are.

A less obvious Castle is Apple, one of the most valuable companies in the world. Clearly Apple is a masterful technology business, but that doesn't make it a new power company. In fact, it deploys an old power model and typically defaults to old power values. It provides highly desirable products to a fanatical consumer base, and does so with a "we know best" ethos. Its product designers in Cupertino, led by the mythical Jony Ive, figure out what we want before we want it and then present their creations to us. Our only job is to consume (even when they decide we no longer need our headphone jacks). There are "open" flanks in Apple's business model, like its app store, but even that is subject to onerous restrictions and centralized control (and extractive behavior) by Apple. Culturally, Apple is known as secretive and an uneasy collaborator. Its continued ascendance is an important reminder of how successful old power models can still be. But most organizations will find it difficult to conjure up Apple's magic.

We finish this tour in the top-left quadrant, with the **Co-opters**. Here we find organizations with new power models that seem to live by old power values. For every Wikipedia—a new power model that remains a beacon of openness and democratization—there's an Uber or Facebook, both of which have reached huge scale through their remarkable peer-driven networks, yet tend to default to old power values in how they relate to their users, hoard value, and share information. We see Co-opters facing increasing pressure as savvy crowds and savvy competitors start to highlight the discrepancy between their utopian messaging and the way they wield power. But in many ways these are also thriving: the Islamic State, white supremacists, and other digitally savvy hate groups deftly combine decentralized social media armies with values that are profoundly authoritarian.

Many of these organizations are moving, or trying to move, around the compass. We see *The Guardian,* for example, trying to shift closer toward the Crowd quadrant with experiments in reader participation. Black Lives Matter began as a purely open-source effort, but as it grows it is beginning to embrace more formal organization, with chapters emerging around the country to consolidate its efforts and reaffirm its original commitment to bringing marginalized voices into the center of the movement.

Many old power companies, from GE to Unilever, have undergone major changes in recent years to align their business practices and internal cultures with aspects of new power thinking. But shifting business models in more fundamental ways is more difficult. Open innovation has often stayed at the margins of corporate transformations, for example. Yet as the book unfolds, we'll tell the stories of old power companies that have made the leap, and show you how. (In the process, you'll learn what an Adult Fan of Lego is—unless, that is, you are one already.)

Similarly, the new power movements and models that find success and amass big crowds face tough choices about whether to maintain their founding commitments to new power values, or even shift away from their new power business models. In a world of old and new power colliding, competing, and converging, everyone is on the move. All organizations need to consider where they are on this compass, where they should move in the coming years, and how they are going to get there.

Whether headed at speed to the top right or working slowly and strategically to break out of the Castle quadrant, we all now need to understand—and be able to deploy—a fresh set of new power skills. Those new capabilities, and their implications for our everyday lives at work, at play, and as a society, are what this book is all about.

FROM SOUND BITES TO MEME DROPS: HOW IDEAS SPREAD

Trust British pub-goers to conjure up a game like Neck and Nominate. Here's how it works. You are filmed drinking a pint of beer in one gulp (hence "neck"), you name someone else to do the same, and you share the video online. Your nominee follows suit. Neck. Nominate. Repeat until drunk.

After circulating among students in 2008, Neck and Nominate really took off in November 2012, when a gentleman named Will Green took things out of the pub and added a very odd twist. Knocking on the door of an unsuspecting neighbor, he waited until she opened up and then necked his beer. The video closes with a woman asking rather politely (and not unreasonably), "Who are you?" as Green departs in triumph.

On the other side of the Atlantic, in March 2014, Jessica Lagle was pushed off a dock by one of her children after she dawdled too much in advance of completing what she called the "24 HR cold water challenge." Before her kids intervened, you could hear her nominate a range of friends to take the challenge. In the video posted on YouTube, she directs people to support a cause she cares about: Marcelly's Dream, an evangelical mission to Africa.

That May, the "cold water challenge" popped up in Lexington, Kentucky. In this iteration, firefighters would douse groups of people with their hoses in exchange for donations to support a

colleague with cancer. The experience became a local hit. By June
the rules of the game had taken shape. A piece on *KYForward,* the
statewide philanthropy blog, explained: (1) After being "called out,"
individuals have twenty-four hours to take the Cold Water Challenge; (2) they must make a $10 donation to the Joe Vissing medical
fund; (3) individuals accepting the challenge must call out three
other people in turn; (4) those who choose not to take the challenge
are encouraged to donate $100 as a forfeit.

After the students, moms, and firefighters came the golfers. The
heat of the summer saw groups of serious golfers start to spread the
game through their networks, adding their own twist. They called
it the "Ice Bucket Challenge" and replaced the pint glass/lake/hose
element with the buckets that typically cool their drinks, using
them to tip water over their heads.

On July 15, the chain arrived in the gloved hands of the semi-
pro Chris Kennedy, who posted his Ice Bucket Challenge with a
new beneficiary, the ALS Association, an organization committed
to eradicating a disease a relative suffered from. This moment has
earned its place in philanthropic history as the first time the Ice
Bucket Challenge and ALS were coupled together. But the true catalytic event occurred two weeks later, when Boston-based baseball
player and ALS sufferer Pete Frates posted his own version, complete with Vanilla Ice soundtrack: "Ice Ice Baby."

You'll remember the rest, and you might even remember being
mildly irritated by having your social media feed utterly dominated
by it that summer. From its launch in Frates's Boston, the challenge
circumnavigated the world, bringing in celebrities, politicians,
sports stars, and everyday people, from Oprah and Mark Zuckerberg to a 102-year-old British great-grandfather, Jack Reynolds, the
oldest person to take the challenge. "It was very chilly—in fact it
was bloody cold, especially just in Union Jack boxer shorts," Reynolds remarked. "But some lovely women with a warm towel and a
shot of Grouse whisky soon warmed me up!" (It seems that to the
Brits, everything is a drinking game.)

Between June 1 and September 1, more than 17 million videos
related to the Ice Bucket Challenge were shared on Facebook alone,

viewed more than 10 billion times by more than 440 million users. The ALS Association ultimately raised $115 million over the summer, more than four times its entire annual budget.

For the ALS Association, 2014 was the year of the "meme heard around the world." But go back just months and the organization was chugging away in relative obscurity. Its annual report noted steady progress, nothing spectacular. The group released a three-part DVD series on respiratory care. Its Veterans Day ALS awareness campaign generated fifty letters in print and online media. ALS Awareness Month delivered an "183% increase" in website traffic. Then, thanks to the iterative efforts of millions, the ALS Association became the unplanned, unlikely recipient of a huge surge of new power.

The Ice Bucket Challenge—love it or hate it—was a phenomenon that tells us something important about our era. By unpacking how and why this campaign went so big, we can learn a lot about how ideas—good, bad, and ugly—spread in a new power world.

THE BIG DIFFERENCE BETWEEN A SOUND BITE AND A MEME DROP

In the old power toolkit, our hammer and wrench were the slogan and the sound bite. As radio, then television, dominated the media, we placed a premium on the perfectly crafted phrase that lodged in people's minds: "We shall fight them on the beaches," "Read my lips: no new taxes," "Just do it," "Mission Accomplished," "Where's the Beef?" These messages were designed for download, made to be distributed to the masses via a limited number of mighty mediators who were the primary link between large audiences and the institutions and brands that sought to reach them.

The striking thing about that era was how much of our cultural experience was shared. Most people would watch the same few television shows and read the same kinds of newspapers. If you had access to mainstream media or could pay to advertise, you were one of the very few who could really shape the culture. Without that access, your ideas would be fringe at best. In such circumstances, old power media companies were able to capture enormous value.

The rise of new media changed things. Organizations and individuals began to bypass big media and tell their own stories. Audiences fragmented. The communications pathway no longer took a mandatory stop at Madison Avenue or the office of the local newspaper. Yet when the medium changed, the messaging didn't always. Organizations relied on old power defaults. Stories continued to be downloaded to audiences, only now they also arrived via the company blog or Twitter feed, rather than simply via a press release or a two-page spread in *Fortune*. "Craft and blast" remained the dominant approach.

In fact, platforms like Facebook and Twitter still largely rely on a classic approach to advertising. Companies pay to intrude on the things you care about (your newsfeed or the video you want to watch) and get you to focus on their product or ideas. Sure, this is sometimes a little sneakier than TV commercials, and certainly a lot more targeted, but the dynamics are similar. Even the much-vaunted "native advertising"—where online advertising messages appear as if they are editorial content—is little more than a fresh spin on an old record, the advertorials and product placements for a new generation.

But there *is* something new happening, which we saw on a grand scale with the Ice Bucket Challenge, and which is reshaping how we think about spreading ideas. The job now is not simply to create sound bites, but what we call "meme drops"—whether images or phrases, across every type of media—that are designed to spread "sideways," coming most alive when remixed, shared, and customized by peer communities, far beyond the control of the meme creator(s). The Ice Bucket Challenge worked not because it was a perfect piece of content, like Nike's "Just Do It" slogan, but because it created a compelling *context* to seed activity by people all around the world. It was a blueprint for action dropped into the fast-moving current of ideas and information, ready to be taken in countless directions, in countless forms.

FROM "MADE TO STICK" TO "MADE TO SPREAD"

In the craft-and-blast era, campaigns were designed to be instantly memorable. In their terrific best-selling book, *Made to Stick,* brothers Chip and Dan Heath unpack the concept of "stickiness," or what it is that makes ideas stick in our brains. They cite many examples, from JFK's "Man on the Moon" speech, to the urban legend about the tourist who wakes up with a kidney missing, to the commercials featuring Jared, who lost half his body weight by eating Subway sandwiches. (An association that Subway now wishes wasn't quite so sticky . . .)

What these examples share are six qualities the Heaths see behind a sticky idea:

Simple—simplicity is the key
Unexpected—surprises you and makes you want to know
  more
Concrete—creates a clear mental picture for people
Credible—uses statistics, expert endorsements, etc.
Emotional—appeals to deep human instincts
Stories—takes you on a journey that helps you see how an
  existing problem might change

The last *S* doesn't stand for anything, but they leave it in anyhow, on the basis that **SUCCES** is not as sticky an acronym as **SUCCESS**. The book is useful for anyone wanting to understand how ideas catch on and resonate in people's minds. These are evergreen principles of making ideas stand out.

But the rise of new power is asking us to consider some new principles, too. We may understand how to "make it stick," but in a world of manic participation, awash in information, how do we "make it spread"? When people are no longer satisfied with simply consuming ideas, but increasingly expect to play a role in developing, tweaking, and propagating those ideas to an unlimited potential audience, what makes a winning idea in the twenty-first century?

With a hat tip to the Heath brothers, we propose that many of the most successful ideas and communications strategies today add **ACE** to **SUCCESS**. **ACE** stands for the three design principles key to making an idea *spread* in a new power world:

- **Actionable**—The idea is designed to make you do something—something more than just admire, remember, and consume. It has a call to action at its heart, beginning with sharing, but often going much further.
- **Connected**—The idea promotes a peer connection with people you care about or share values with. Connected ideas bring you closer to other people and make you (feel) part of a like-minded community. This sets off a network effect that spreads the idea further.
- **Extensible**—The idea can be easily customized, remixed, and shaped by the participant. It is structured with a common stem that encourages its communities to alter and extend it.

We can see these qualities playing out in the Ice Bucket Challenge.

It was **Actionable** in several clear ways. Of course, it asked you to make a donation. But that wasn't the central action. It wasn't even the primary result. Many more people took part in the challenge than donated to ALS (this was a widespread and well-made criticism of the effort). It asked you to create a video and to share it. It asked you to nominate your friends to participate. It asked you to like, share, and comment on other people's posts.

It was **Connected** on at least three levels. First—and most important, perhaps—it connected you to your own immediate peer group via sharing and nominating. The initiative gave people what we call *permission to promote,* and not just for ordinary people but also for celebrities. The Ice Bucket Challenge proved a perfect way to humanize the serious and powerful, like Bill Gates, to their audiences. It was an excuse for young YouTube stars to show off for their fans in a bikini or swim trunks. Second, by linking ordinary people to the famous people taking part, you could share a cause and a moment with everyone from Lady Gaga in her leather bondage suit to LeBron James in his skivvies on the deck of a yacht. Third, it invited you to join a new global affinity group that coalesced around supporting ALS. It asked you to contribute to something much bigger than you. The "good cause" factor of ALS was the backing track. But you were the rock star, your friends were your rhythm section, and your social network was your audience.

It was **Extensible** in the sense that every Ice Bucket Challenge action was unique, each video customized and personalized for its own audience. This was a world away from the craft-and-blast strategies of old. The potential for small tweaks and personal flavors invited everyone to become not simply a participant, but a producer. Manju Latha Kalanidhi, a journalist from Hyderabad, India, turned the Ice Bucket Challenge into the *Rice* Bucket Challenge. Patrick Stewart of *Star Trek* fame shared his own elegant take in a wordless video of him writing a check, then fastidiously extracting two ice cubes from an ice bucket with a pair of tongs, pouring some single-malt whisky into a tumbler, and taking a slow sip. The Ice Bucket Challenge mixed that perfect cocktail of emotion: it made people feel the same, in that they were all part of a communal effort, but it also made people feel different, with each video feeding the agency of its creator and network.

These three principles—**Actionable, Connected,** and **Extensible**—can be seen playing out in many of the successful efforts to spread ideas in recent years—be they social movements like #MeToo, start-ups, brand ventures, ad campaigns, or even (as we shall see at the end of this chapter) terrorism.

A is for Actionable: What makes BuzzFeed different

By now, we all know BuzzFeed. Some of us are among the 14 million who took its quiz "What City Should You *Actually* Live In?" Others are in the millions who squabbled over whether a certain dress was gold or blue. Many have shared a list like "17 Times Double-Jointed People Took It Way Too Far" or the "100 Most Australian Words of All Time." The hungry may be part of Tasty, BuzzFeed's Facebook community for foodies, 85 million strong. The creative might have put together their own BuzzFeed-style listicle using the site's community-created content platform. And some will even have read BuzzFeed's award-winning reporting on everything from presidential politics to the lives of trans women: the success of their lightweight content has helped enable a more serious investment in heavyweight journalism.

BuzzFeed began as a joke to many—especially the establishment. But no one laughs anymore. It has been valued north of $1.5 billion and named one of the Most Innovative Companies by *Fast Company*, which billed it the "envy of the media world."

At the core of its success is what it wants its reader to *do*. Its primary goal is not that its content be read, but that it be *shared*. As Ben Smith, editor in chief, puts it:

"If your goal—as is ours at BuzzFeed—is to deliver the reader something so new, funny, revelatory, or delightful that they feel compelled to share it, you have to do work that delivers on the headline's promise, and more. This is a very high bar. It's one thing to enjoy reading something, and quite another to make the active choice to share it with your friends. This is a core fact of sharing and the social web of Facebook, Twitter, Pinterest, and other platforms."

It is easy to think of this as trivial behavior, and in many ways it is. But BuzzFeed has triumphed by taking it so seriously. Its data-driven and analytical approach to understanding its audience, what makes them tick and how to activate them, is key to its success.

Old media is cottoning on. In 2013, the single most popular piece of content produced by the *New York Times* was not a groundbreaking piece of investigative journalism, but a twenty-five-question quiz that allowed readers to find their "personal dialect map," the place in the United States where people used language and phrases most similarly to them.

Compare BuzzFeed's drive toward action with the philosophy of *The Economist,* one of the oldest and most respected media brands in the world. As Paul Rossi, president of the Economist Group media businesses, has said, "We want to create value in your head." In doing so, they have achieved huge influence: the Economist Group produces content that heads of state and captains of industry rely on weekly. It would not be caught dead running a piece like "If None of These Pictures Make You Say 'What the Fuck,' Nothing Will."

Yet it is hard to imagine that institutions like *The Economist* will not need to master some of the skills that have made BuzzFeed so successful. It was founded as a campaigning organization, launched to repeal the British Corn Laws in 1843. It has been out in front on some of the biggest cultural and political issues of the day. For instance, in the mid-1990s it made a principled endorsement of same-sex marriage, an idea its editors conceded at the time was considered "strange and radical," helping the idea gain further mainstream legitimacy.

To retain this important role at a time when facts themselves are being brazenly challenged in our public debate, institutions like *The Economist* must be serious about engaging their community around the principles they cherish. (Later in this book, we'll tell the story of the Dutch media company De Correspondent, which offers a master class in how a media company grounded in serious journalism can mobilize its readership into a community.)

Creating Actionable ideas is about more than trolling for "likes." It asks an organization to consider how action by its community can be embedded in the very structure of its communications. It is a philosophy, not a technology, with the core principle that your community members are there to do more than simply consume or comply.

C is for Connected: What we can learn from
Facebook's first effort at election-tinkering

When Facebook was accused of influencing the outcome of the 2016 presidential election, it was not the first time its algorithms had shaped political destinies.

Back in 2010, Facebook launched an experiment to test whether its platform could contribute to political mobilization, sending out 61 million messages in the lead-up to the U.S. congressional elections. It divided those who received the messages into three groups.

One group got an "informational message" that included information about where they could vote, encouragement to do so, and an "I voted" button they could press. Let's think of these people as the "billboard" group.

Another group got no message at all. Let's call these folks the control group.

Most people received a "social" message, which was the same as the "billboard" message but with one big difference. It showed the profile pictures of up to six randomly selected Facebook friends who had clicked the "I voted" button.

Researchers from the University of California, San Diego, in collaboration with Facebook's data-science team, then compared online actions with public records to get a sense of whether which message the user received (or did not receive) affected whether the person voted. They published their study in *Nature*.

Their first stunning result was that the billboard group voted at the same rate as the control group. Even with the benefit of what many would think of as the most valuable digital advertising space in the world (the top of your newsfeed) and an actionable "I voted" button, voter turnout among this group was not markedly different from people who received no message at all. This has profound implications for those organizations that operate in soundbite mode. Capturing people's attention can only get you so far.

The second conclusion came from the difference in mobilization between the "billboard" group and the "social" group. The simple addition of pictures of users' friends who had voted led to very different results. Members of the "social group" were 2 percent more likely to click the "I voted" button, and the researchers estimate that the social message increased turnout by 60,000 people. This was the first study to demonstrate that the online world can "affect a significant real-world behavior on a large scale."

The third conclusion came from studying people who weren't part of any of these groups, but saw posts come up in their news-feeds as their friends pressed the "I voted" button. Researchers found that 280,000 of these "indirectly" influenced users turned out to vote. This group were *four times* more likely to vote than those who saw nothing.

The scientists concluded that the Facebook experiment led to a non-trivial increase in voter turnout, the kind that might be critical in a close election. And those people turned out not because the candidates had compelled them to, or because the ad at the top of their newsfeed exhorted them to, but because it is what they saw their friends doing.

When we think about how ideas spread in the twenty-first century, the experiment makes it clear that they spread *sideways*. A big idea is made stronger if it is designed to pass between small groups of friends. MIT's Alex "Sandy" Pentland, who detailed the Facebook voting experiment in his book *Social Physics,* told us: "People actually begin changing behavior when [an] idea gets validated by their community, rarely when it has not."

This is why "connected" ideas matter so much. Today, the most resonant ideas are not those that get flashed at the highest number of people but those that become individualized expressions of affiliation and identity among peers. As former BuzzFeed president Jon Steinberg and Jack Krawczyk noted in *AdAge*, "Our data show that online sharing, even at viral scale, takes place through many small groups, not via the single status post or tweet of a few influencers. While influential people may be able to reach a wide audience, their impact is short-lived. Content goes viral when it spreads beyond a particular sphere of influence and spreads across the social web via . . . people sharing with their friends." The good news here? You probably have much more influence over your friends than Kim Kardashian does.

Connectivity can be supercharged when it links someone, along with her peers, to a larger belief or philosophy. This effect explains the wild popularity of *Humans of New York,* a blog turned best-selling book by the street photographer Brandon Stanton. Beginning in 2010, Stanton posted candid portraits of everyday people, along with a brief highlight from short interviews he conducted with his subjects. The pictures range from the commonplace—a mother and son on a bench with the caption "We're talking about how important it is to take a shower every day"—to the heart-wrenching: a photo of a woman looking troubled over the words, "I live with my mom and uncles. There are twelve of us in one house. Everyone is sacrificing so I can go to university."

Humans reaffirms a belief that stands out in a world of shiny celebrity: normal people matter. Stanton's portraits are love letters to the ordinary.

These love letters get shared a lot. Stanton has 17 million Facebook followers who have over time built a sense of community with each other. Each picture he serves up offers them something of real value, a daily demonstration of our common humanity. The like or comment or share his followers add takes them just seconds, but it is a powerful way for them to signal to their peers "we are people who care."

E is for Extensible: How #GivingTuesday turned into #GivingBlueDay . . . and #GivingShoesDay . . . and #GivingZooDay . . . and #UnDiaParaDar

In 2012, a new idea came out of an old institution. Henry and his colleagues at the 92nd Street Y ("92Y")—a 144-year-old New York City cultural institution with hundreds of employees running thousands of local programs—launched #GivingTuesday, a campaign designed to encourage a day of charitable giving after the retail bonanzas of Black Friday and Cyber Monday. The goal was to create an "opening day" for the giving season: a day of altruism after all that consumption. It was a classic "meme drop": Henry and his team seeded the idea in the world, built a website offering some tools and tips for organizations that wanted to run campaigns, and then largely let the crowd run with it.

In its first five years, it has inspired blood drives, coat drives, volunteer projects, and civic renewal efforts, as well as generating hundreds of millions of dollars of donations for nonprofits. The 2015 event saw PayPal set the Guinness World Record for most money raised online in twenty-four hours. Micah Sifry, the civic technology thinker and co-founder of Civic Hall, described it like this: "Arguably, it's become the most successful civic tech culture hack of the decade."

The old power approach to an idea like this would have required a high degree of control and ownership by 92Y. There would be one way to take part and everyone would have to ensure 92Y was properly credited in every mention of the campaign.

But this was a new power idea, intended to be extensible: adopted and then mutated by those who carried it forward. So when the University of Michigan decided to turn #GivingTuesday into #GivingBlueDay, in an effort to make it more meaningful to its alums, 92Y saw it as a sign of success. In the old power world, such an appropriation would be met by a call to the lawyers. "Cease and desist," they would exclaim. "Our brand integrity is being threatened!" But the 92Y team knew that by adapting #GivingTuesday, the university had increased its ownership stake. It was a boost to the project, not a threat.

This approach paid off. #GivingBlueDay was the first online fundraising day the university had ever run. It set an ambitious goal of $1 million—and raised $3.2 million in year 1, $4.3 million in year 2, and $5.4 million in year 3. And Michigan was not alone in tweaking the brand to make it speak more clearly to its cause. The organization Dress for Success turned #GivingTuesday into #GivingShoesDay, and now receives donations of thousands of pairs of shoes for underprivileged women re-entering the workplace. There is a #GivingZooDay where zoos across the United States have united to showcase how much they contribute to their local communities.

Things got even more interesting when people changed the frame altogether—yet kept the spirit. In Baltimore, the community came together to launch #BMoreGivesMore, a #GivingTuesday campaign with the goal of showing off the city as "the most generous in the USA." Everyone from local pizza parlors to big businesses played their part. Together they raised over $5 million for local charities.

On the other side of the world, leaders in Sorocaba, a big Brazilian city, heard about the success of the Baltimore campaign and developed their own campaign, which they called "Doa Sorocaba." Along with cash donations, it brought the community into the streets to donate blood, food, books, and more, replicating Baltimore's success with a Brazilian twist.

As the movement has spread globally—it is now in almost a hundred countries—the frame of the idea has continued to expand. In South America, #GivingTuesday is #UnDiaParaDar. Singapore has changed #GivingTuesday into #GivingWeek. In Russia it is #щедрыйвторник.

All of these extensions make the movement stronger. Around the world, people take action at the same time and with the same intention, but the concept can be adapted to fit the local culture. This drives ownership and resonance in a way a strict franchise-style model never could.

Often the most extensible ideas feel imperfect and incomplete; if the idea feels "untouchable" or overly polished, it is very hard for others to feel they can take the reins and make it their own.

An award-winning example of an extensible idea comes from Taco Bell, which for Cinco de Mayo in 2016 created a special lens on Snapchat that allowed people to turn their heads into giant taco shells and have hot sauce poured over the top. It won the title of most popular lens in Snapchat history by scoring 224 million views in one day.

Many will scoff, but contrast this engagement to the dynamics of taking a traditional ad slot during prime-time TV programming.

The guy on his couch, eating potato chips, may or may not pay attention to the message blaring out of the TV at him and everyone else in that cable district as he waits impatiently for his favorite show to start up again. But the taco-headed Snapchatter is actively engaged with the brand. *Adweek* reported that "the average user played with Taco Bell's ad for 24 seconds before sending it as a snap." Most important, the user spreads the brand's message to his friends as a trusted and influential voice.

For an ACE idea, going viral is not just an explosion of sharing. It is allowing every carrier of the message to add a mutation that makes it more contagious for his or her own networks. People talk a lot about movements these days, but it's only really a movement if it moves without you.

THE BATTLE FOR IDEAS: ARE YOU THINKING ICE BUCKET OR TELETHON?

An unintended consequence of the Ice Bucket Challenge was the effect it had on its old power predecessor, the telethon. Started in 1956, the Muscular Dystrophy Association (MDA) telethon was begun by Dean Martin and Jerry Lewis, but it was finished off by millions of people worldwide, who were no longer satisfied with a Pavlovian response to the celebrity ask.

"The decision to end our beloved telethon was not made lightly," said Steven M. Derks, former MDA president and CEO. "In the last few years, the show was adjusted to reflect changes in viewership and donor patterns, and last summer's Ice Bucket Challenge once again affirmed for us that today's families, donors and sponsors are looking to us for new, creative and organic ways to support our mission."

The telethon model was actionable on a very basic level—it implored people to call in—but it wasn't especially connected, and it wasn't at all extensible. Dialing in to donate in the hope that a celebrity might answer the call did not offer a meaningful enough connection, and lacked a peer-to-peer element. And there was only that one way to be involved.

However, for all the old power design features of the telethon, it certainly delivered. For over sixty years, it brought regular and reliable funding and awareness to the Muscular Dystrophy Association. The Ice Bucket Challenge, by contrast, for all its "new, creative and organic" elements so absent in the telethon model, has proved a lot less reliable. Fundraising for the ALS Association on its second go-round with the Ice Bucket Challenge went from a cascade—$79.7 million in August 2014—to barely a drip—$500,000 in August 2015.

Every organization whose boss or board has laid down the order to its staff "We want our own Ice Bucket Challenge!" has missed the point. What leaders can do is blend ACE principles into all of their efforts, and build the capacity to do this work consistently, rather than praying for the occasional freak deluge.

THE ACE IN ISIS . . .

"Sometimes, it's better to just leave it all, and just breathe."

The reassuring white block text sits superimposed on a filtered photo of a plane taking off into the sunset, one of several memes Umm Layth has posted to her popular Tumblr blog.

It's a kind of post familiar to anyone who's casually browsed a blog. But unlike most young women curating vague and wistful memes on their social media accounts, Umm Layth is being quite literal. Amid her goofy emojis, colorful street photos, and GIFs, Umm Layth is above all imploring her readers to pack their things, get on a plane, and take off—in the name of joining the Islamic State. Her success is a sobering reminder that ACE ideas aren't just influencing how we raise money or move products.

We met Umm Layth at the very start of this book. "Umm Layth" is widely understood to be the internet moniker of Aqsa Mahmood, a young Scottish woman who left her family home in Glasgow in November 2013, at the age of nineteen. Friendly faced and educated in good schools, her parents never imagined the possibility she would make *hijrah* to the Islamic State in Syria, a topic her Tumblr blog visits over and over again. The term, she explains, means "to separate [from] or abandon" a non-Muslim place for "a place where there is a presence of Islam." In August 2014, she reblogged a tweet from @Bintladen: "When you separate from your Hijrah buddy for the first time in 8 months." It's capped with a frowning emoji.

Mahmood has been called an "ISIS poster girl," and her popular Tumblr blog and Twitter account (now banned) evince a mastery of twenty-first-century communications tools and tactics. And she's hardly the only one—there's a whole subset of so-called "jihadi brides" who regularly offer advice and guidance to other potential girl recruits by skillfully using the conventions of Tumblr and Twitter, even including sample packing lists for making *hijrah*, down to the smallest of details.

Over Twitter, she has offered personal encouragement to fighters: "Follow the example of your Brothers in Woolwich, Texas and Boston, etc. Have no fear as Allah . . . is always with the believers."

Mahmood and others also use an encrypted online messaging app called Surespot as a way to groom potential recruits in the United Kingdom and, as the *Daily Record* reported, "to send secret advice from Syria."

By communicating directly with their peers, Mahmood and others like her help ISIS cultivate a community, a sense that those who answer its violent call will find belonging and support among a new family. As for the families they leave behind, they do not go totally forgotten. Mahmood, in what seems a highly strategic show of empathy with those she seeks to recruit, writes emotionally resonant meditations on how hard it has been to leave her family.

"The first phone call you make once you cross the borders is one of the most difficult things you will ever have to do," Mahmood tells her followers, in a post that's been reblogged and liked dozens of times. "When you hear them sob and beg like crazy on the phone for you to come back it's so hard."

Over the past several years the Islamic State has skillfully used many of the principles we've described in this chapter to spread its ideology of hate and violence—and recruit tens of thousands of women and men to its cause. It thrives on a highly effective combination of cruelty and coercion—waging war, kidnapping, acts of terror—with a "made to spread" ACE communications strategy that has inspired many acts of violence by people it has neither trained nor directed. There is perhaps no organization in recent years that has spread ideas so far, and so unexpectedly, as ISIS.

The Islamic State may want to take us back to the Dark Ages, but its propaganda machine is multi-platform, multi-pronged, and distinctly modern. It has invested in old power strategies such as the creation of centralized media outlets like the Amaq news agency that pump out content and broadcast messages from its leaders. And it has created a decentralized "content strategy" perfectly adapted for the social media age. It has tweaked familiar memes, blaring "#YODO—You only die once. Why not make it martyrdom?" and encouraged fighters to post photos of their bloodied hands after their first kills (a horrifying example of an extensible idea). The effect of all this engagement is chilling. The group's presence online is so strong that it has created a self-fulfilling dynamic, inflating its perceived power and overshadowing other extremist groups with which it competes for recruits.

The Guardian reported in 2014 that much of this online activity isn't generated from ISIS "headquarters" but is driven by admirers churning out content without any initiation or direction from above. At that time, social media postings helped to drive mass Iraqi army desertions because soldiers feared that the Islamic State was about to take Baghdad. As ISIS supporter Abu Bakr al-Janabi put it: "The picture that said 'Baghdad, we are coming'—nobody asked [its creators] to do it, but they did it anyway." As The Guardian commented in 2014, "Now, in the age of YouTube, Twitter, smartphones, cheap cameras and software, the superpowers no longer control information. Ironically, the beneficiaries of this media democratization are a medieval theocracy hell-bent on eradicating democracy from the face of the earth."

The Islamic State has wreaked havoc by making its supporters the owners and shapers of its ideas.

HOW TO FIGHT BACK: ACE VS. ACE

After its airdrop leafletting strategy proved ineffective, the U.S. government tried to get serious about using social media. Its Center for Strategic Counterterrorism Communications (CSCC), in an effort to target those sympathetic to jihad but who have not yet made the decision to turn to violence, launched its "Think Again Turn Away!" campaign. Its goal was to provide a counternarrative for wavering recruits.

On the surface this seemed a promising idea. But the reality, as we described in chapter 1, verged on self-parody: a profile picture featuring the insignia of the U.S. State Department and a Twitter banner featuring a picture of the State Department's dour concrete office building. Their Twitter account was snarky and kept picking online fights with jihadis. The jihadis often got the better of these tussles.

Perhaps the lowest point was the release of the State Department's "comic" video, "Welcome to the 'Islamic State' Land (ISIS/ISIL)," which tried to combat extremism through the power of satire (sample line: "Travel is inexpensive because you won't need a return ticket"). This attracted derision not just from those it was trying to convert, but from late-night comedians who lampooned the lameness of the efforts.

After these false starts, the United States began to realize it needed to take a different approach. CSCC coordinator Alberto Fernandez laid out a new direction, speaking to Congress in 2015: "You need to find a way to form loose, open source communities of interest or swarms that can swarm back and push back against the ISIS message. It's not an impossible thing to do. It can be done."

The new interagency Global Engagement Center is trying to make this happen. Discarding the top-down hectoring tone of the Think Again Turn Away campaign, the center is trying to build a "network of positive messengers" to share not just counternarratives, but *alternative* narratives drawing people away from more extreme positions, amplifying the messages of its partners, from religious leaders to schools.

One promising effort is the P2P (Peer to Peer) Challenging Extremism competition, which partners with Facebook and hundreds of universities around the world. The brief here is loose: students come up with creative ways to "push back on online hate, prejudice and extremism while empowering their peers."

A winning entry was FATE—From Apathy to Empathy—a campaign created by students from the Lahore University of Management Science in Pakistan to fight the desensitization that comes from constant exposure to extreme images, as well as to counter stereotypes about Muslims. It mobilized the masses to tell stories of how terrorism had affected their own lives.

Another came from a group of Finnish students who launched a movement of pop-up restaurants where asylum seekers shared their native foods with locals. A U.S. team created a Snapchat campaign. A class from Azerbaijan created a tolerance toolkit for teachers to use in the classroom.

Of course, there is no "killer app" for defeating extremist propaganda, no one idea or moment that will suddenly change the game. Tech platforms need to do much more to halt the spread of extremist ideas, something that is becoming a greater commercial concern as advertisers pull their spending from platforms like Google and Facebook out of worry their ads will appear next to extremist content and fake news. For too long, their business models have benefited from the fact that provocative content of all kinds is often more seductive and popular than the bland and virtuous. (As one promising response to the problem of extreme content, Google's think-tank Jigsaw launched something called the "redirect" method, which serves potential recruits video testimonials from credible voices and other content designed to dissuade them when they search for a range of keywords typically used by wannabe ISIS members.)

The United States and its allies need a decentralized social media army of credible figures—moderate Muslims and others, who probably aren't going to be big fans of U.S. foreign policy—who can make a convincing peer-to-peer case for a different path. In a new power world, this battle of ideas is not a standoff between bureaucrats and terrorists. It is a showdown between Scottish teenagers and Pakistani business students.

The future will be won by those who can spread their ideas better, faster, and more durably. In a world of fake news, climate change deniers, Holocaust deniers, anti-vaxers, and all manner of extremists, the stakes are high. Those on the side of the angels, who want to spread compassion, promote pluralism, or defend science, must first grapple with a painful reality: that new power can supercharge hate and misinformation. In fact, those darker forces often start at an advantage because their provocations compel our attention and our clicks. It isn't enough to simply have the facts on your side. Truth needs ACE.

HOW TO BUILD A CROWD

The protest begins with an ethnically diverse crowd of the most well-styled, shabby-chic young people you have ever seen. They march down the street smiling, pumping their fists in the air, holding up signs with the oddly generic message "Join the Conversation."

As things unfold, you can see people around them going on with their lives: a pair of women are brunching; a cellist practices in an industrial-looking space; a young woman wearing a hijab looks like she is struggling with a photography project; and there is Kendall Jenner, reality star and member of the Kardashian clan, pouting at a fashion shoot. An anthem by Bob Marley's son Skip blares in the background: "We are the movement! This generation!" You notice that everyone has something else in common: they are all drinking Pepsi.

You may already know what happens next: the cellist, the young Muslim photographer, and Jenner herself join the crowd. Kendall rips off her blond wig in an attempt at solidarity and approaches a line of stern-looking police officers. Anointing herself the leader of the protest, Jenner—a very attractive white woman—courageously walks toward an equally attractive white male police officer and hands him a Pepsi. The officer takes the Pepsi, cracks a smile in return, and the threat of violence is lifted. The protesters erupt in jubilation.

This ad for Pepsi was an epic debacle. With the Black Lives Matter movement and the anti-Trump "resistance" in full swing in early 2017, Pepsi clumsily stole from the cultural zeitgeist. This *did* inspire a movement: it unified all sides of the political spectrum against Pepsi. As *New York* magazine put it: "Pepsi's Kendall Jenner Ad Sparks Mass Mobilization of Mean, Funny Tweets." Under huge pressure, Pepsi pulled the spot (and, rather oddly, apologized to Kendall Jenner).

The Pepsi ad spoke to a world where movements are proliferating and going "mainstream." Having a connected and passionate crowd on your side has become a crucial asset, whether you're trying to get a law changed or an app launched or just trying to sell soda. Movements, both real and Pepsi-real, are popping up everywhere, originated by brands, organizations, and even teenagers who run multimillion-person communities from their bedrooms.

In this chapter, we explore how crowds are built in a new power world, and share the most important lessons we've learned as movement builders. We start by going back in time a little, to look at one of the first movements Jeremy ever helped create.

THE STORY OF GETUP

It was late 2004. John Howard, the prime minister who had run Australia since the mid-1990s, had just been elected to a fourth term in office. Many Australians, including Jeremy, were despondent. Howard had led Australia into the Iraq War, refused to apologize for the historic injustices committed against Australia's Indigenous people, and championed a policy that led to thousands of refugees, including children, being locked up in desert prisons.

Not everyone shared Jeremy's view. Howard appealed to a strong traditionalist and nativist strand in Australian culture and was a very savvy politician. After four straight election victories, the pundits said he was unbeatable. The opposition Labor Party was in disarray, and the many people who felt ill at ease with Howard's

1950s-era, white-bread vision of Australia were at the point of giving up.

One day, just after Christmas, Jeremy sat on Sydney's Bondi
Beach with his friend David Madden and struck up a plan. David
and Jeremy had been to graduate school together at the Harvard
Kennedy School and had just returned from a year in the United
States volunteering on a campaign against the Iraq War. During
that year, they'd begun to experiment with the kinds of new power
tools like digital activism and crowdfunding that would come to
transform politics in the United States.

They had both taken part in big protests in Australia over the
years, such as joining hundreds of thousands walking across Sydney's iconic Harbour Bridge calling for reconciliation with Australia's Indigenous people. Yet, for all the energy of these moments,
people went home and things tended to pretty quickly go back to
normal. The big organizing challenge was not just to rally people for
a single passionate protest, but to find a way to channel that passion
again and again.

This was an ambitious goal for two people in their twenties who
had never been members of any political party and had no formal
political connections, but they had the great advantage of being too
young to know any better.

So in August of 2005, Jeremy and David, together with their
friend Amanda Tattersall (who had helped to introduce the idea
to existing progressive and labor groups), launched GetUp—an
effort to bring participation back into democracy. They made a TV
ad boldly declaring that Australians were "starting a movement"
(though at this point the movement was little more than Jeremy's
older sister and his seven-year-old niece, the stars of the ad). But
from the start, they believed that Australians would rise up if given
a little more hope and a lot more agency. Jeremy's sister's big line
in the commercial summed up the behavior they hoped to inspire:
"I'm not going to give up, I'm going to get up!"

GetUp's launch asked people to take a simple action: Write
online to your politicians. Its online tool—at the time something of

a novelty—made it easy for ordinary Australians to email their representatives (easier than sending snail mail, which at that time was the primary way people could contact their representatives). It even helped people who did not know who their representatives were to find them easily, removing an important point of delay.

The response was extraordinary. Within days, tens of thousands of Australians used the tool to write to their representatives about the issues they cared about and join GetUp. But what happened next was even more important. The Howard government, and its allies in Australia's Rupert Murdoch–owned newspapers, freaked out. Andrew Robb, a key government politician, appeared on national television to condemn the campaign. "There are hundreds of emails arriving in Senators' offices. They're beside themselves, just to clear the screen," he said. "This is highly irresponsible, this is spam, this is spam."

If you know anything about Australians, you will know that being told that their personal messages to their elected representatives were "spam" was not likely to delight them. And it got worse. The government actually launched multiple investigations into GetUp to try to shut it down, referring it to the Electoral Commission and even, improbably, to the anti-trust authorities. All this pushback, ironically, supercharged GetUp's credibility. People started to believe that the organization might actually be a threat to the established political order. Online donations surged. New members flooded in. A movement was born.

GetUp would go on to play a part in delivering what just two years before had been a very unexpected outcome. By the time of the 2007 federal election, the movement was already so big that its membership had thousands of members in John Howard's own electoral constituency. When the votes for Parliament were counted, not only did Howard's government lose power, he was the first sitting prime minister in almost a hundred years to lose his seat in Parliament. Thanks to a campaign masterminded by GetUp's first executive director, Brett Solomon, GetUp members volunteered in unprecedented numbers in Howard's electorate (district), knocking on doors and making the case for change, making 187,000 voter contacts in the lead-up to, and on, election day. The movement for progressive politics in Australia had been born again.

In the years since it was launched, GetUp has influenced the outcome of several additional key elections, won a landmark ruling in Australia's highest court that established new constitutional voting rights, and helped to pass legislation protecting refugees and the environment. It has raised tens of millions of dollars in small donations from its members to fund its work. It's been described as "probably [Australia's] most important civil society actor." Lamenting the disruptive impact the organization has had on the established political order, the conservative commentator Nick Cater wrote that GetUp "is to politics what Uber is to taxis." Today, GetUp is the largest political organization in Australia, with more members than all of the country's political parties *put together*. From its start with two young people on a beach, now more than a million Australians call themselves GetUp members.

THE FIVE STEPS TO BUILDING A NEW POWER CROWD

GetUp is a study in how to build a crowd in a new power world— and how to sustain that power for more than a decade. When GetUp launched, it was an early experiment in digital activism. Today, the tools and tactics available to movement builders have expanded hugely. At the same time, it has become harder to break through because these days everyone is trying to rally some kind of crowd. Whether you're working to get elected to your local school board, launch an online community, or just trying to build buzz around your new business, these are the five key steps to starting and growing a flourishing movement today.

Step 1: Find your connected connectors

In the early days of GetUp, Jeremy and his colleagues weren't trying to rally *everyone*. They were laser-focused on a group with a very specific worldview: people who were compassionate and concerned about fairness and who had a less individualistic view of society and the economy. They were educated, skewed female, and often baby boomers.

GetUp designed campaigns that, time and again, hit the hot-button issues for this group—from protecting their beloved public broadcast network from funding cuts, to advocating for refugees, to taking action on climate change. While on their face these issues were disconnected, GetUp understood there was a high chance that someone who was concerned about climate change also cared about public broadcasting. This group wasn't just philosophically connected; it was socially connected, clustered in big cities filled with urban professionals. With every early campaign GetUp ran, supporters would be encouraged to "tell a friend," and that person would turn into a recruiter, bringing their own—similarly minded—contacts into the community.

Once GetUp got off the ground, it widened its focus to include a much more diverse range of constituents, especially in terms of rural representation. Yet its initial energy relied on a very specific and passionate slice of society who spread its message sideways to hundreds of thousands of Australians.

Think of this group as the "connected connectors," those people who share a worldview, are networked to one another, *and* are influential in their reach. For any new power movement, identifying and cultivating the right connected connectors is often the difference between takeoff and fizzle.

To take a very different example, in 2016 Donald Trump's connected connectors were an overlapping cadre of online activists and meme creators associated with the so-called alt-right, plus the vestiges of the Tea Party and gun rights diehards. This committed base was the catalyst for his crowdbuilding when he ran for president. They shared a broad set of passionate beliefs and were highly connected on message boards and platforms like Reddit and 4chan. They proved extremely effective at spreading messages and memes digitally, disseminating buzz about Trump and often false rumors or negative information about Hillary Clinton.

Etsy, too, owes its early growth to a core of connected users. The online crafting marketplace now has tens of millions of members and generates hundreds of millions of dollars each year. Yet it only got off the ground thanks to small groups of digitally savvy feminist crafters.

The platform was founded in Brooklyn in 2005 by four men. One of them, Rob Kalin, was a quintessential "maker" and wanted a market to sell his crafty, wood-covered computers. He wasn't alone. In the early 2000s there was a resurgence of enthusiasm for DIY artisanship, expressed online on blogging platforms like LiveJournal and off-line at venues like artisan flea markets in culturally progressive cities.

Some of the most passionate members of this community were women with a feminist outlook, opposed to mass-produced corporate capitalism, and looking to forge new kinds of social and economic activity, explained Morgan Brown in his definitive account of Etsy's rise. So the early Etsy team deliberately recruited the most influential crafters at flea markets in Brooklyn and elsewhere, and promised them a place to sell their wares online. Critically, Etsy also provided moderated community forums on its website that allowed these crafters—who already shared a worldview and had off-line connections to one another—to find one another there. Brown cites a blog post from early 2008 that sums up Etsy's efforts to appeal to community:

> Etsy's core mission is to help artists and crafters make a living from what they make. This may seem like an innocuous enough statement, but truth be told, the socio-political-cultural-economic state in which we live makes this a rather bold rallying cry. We want to make history's "way of doing things" undergo a change . . . Etsy is part of a larger move-

ment and we at Etsy want to learn more about the conscious-
ness of feminist crafters in our midst.

This is most definitely not the kind of language you would have
heard at the time from Amazon or eBay. Etsy had found its connected connectors. These women fueled Etsy's growth organically.
There was a virtuous cycle between feminist crafters promoting Etsy
as a platform and promoting their own wares online. New power
was Etsy's sales *and* marketing engine.

Step 2: Build a new power brand

Every company or institution must make key early decisions about
how it will project outward into the world. It has to come up with a
name and settle on a visual aesthetic and maybe some iconography;
it needs to refine a "voice" for how it speaks to consumers or clients.
These things define its "brand"—the way a product or organization
looks, sounds, and feels in the world. The same questions are vitally
important when launching a new power community, but the building blocks of a new power brand are very different from those of a
purely commercial or transactional brand, or a top-down organizational one.

Think of an Amex Black Card or a Bentley, which project luxury and exclusivity, or the seal of the president of the United States,
which is designed to project authority, pomp, and prestige. None of
these gives an ordinary person any agency. But if you are designing
to inspire participation, not simply drive consumption, admiration,
or fear, there is a different task at hand.

The GetUp brand was designed with new power instincts in
mind. The name itself invoked the ability of everyday people to stop
complaining and start reclaiming their own power. At the heart of
its promise was an exhortation for people to act.

Bright orange was chosen as the brand color and an exclamation mark was slapped after GetUp in the logo to make it feel more
like a call to action (noting, of course, that adding an *!* to the end of

your logo does not a successful movement make, as Jeb! Bush can tell you).

The language of GetUp shared that spirit of group action. The "About Us" section of the website was not written in the voice of an organization talking to its supporters or subjects, but in the voice of the members. The team thought of the GetUp brand less as an organization and more as an individual with a consistent and distinctive personality—a compassionate, smart, idealistic, but not too earnest person. They also thought about who this person *wasn't*—a political hack, a know-it-all, an angry fringe-dwelling ideologue.

When the team noticed people showing up to GetUp protests with their dogs dressed in bright orange GetUp T-shirts, and when the phrase "I am GetUp" surfaced on social media, it was clear that the identity and voice of GetUp had aligned with—in fact had merged with—the identity of its members. As we have said, one of the key signs that you have launched a successful movement is that it starts moving without you.

These ideas don't just apply to political movements: Airbnb's brand is based on similar principles. When it emerged in 2008, its young founders, Brian Chesky, Joe Gebbia, and Nathan Blecharczyk, weren't thinking much about new power. They just wanted to find a way to pay the rent on their San Francisco apartment. But as the company grew, it did so less like a franchise, with every room looking the same, and more like a movement, with early users sharing a passion for a new kind of stay—one that provided instant community in a new place, a host to guide them in a new city, a real person to have a cup of tea with before bed.

In 2014, eight years later, Airbnb had grown exponentially, and it had in many ways moved beyond its earlier, more intimate "homemade" feel. But its founders wanted to ensure that the brand remained connected to that original experience. They needed to maintain that critical point of difference from checking in and out of a Best Western; losing the communal and community spirit of their early days was a real business threat. And there was another threat, too: the regulatory challenges to Airbnb that were popping up in cities all over the world. Airbnb had begun to rally its hosts as a way of fighting back against city governments, which made users' bond with the platform even more critical.

So Airbnb relaunched its brand, with a brand story made for the age of new power. Douglas Atkin, an Airbnb executive with the unusual corporate title "Global Head of Community," summed it up as "creating a world where anyone can belong anywhere."

Airbnb's new logo was not designed to be admired, but to be remixed and adapted by different affinity groups within the Airbnb community. The soft, malleable, inverted heart (or pretzel, depending on your perspective) drew a level of online engagement perhaps best summed up by *Fast Company*'s headline: "This Tumblr shows everything Airbnb's new logo looks like, in addition to a vagina."

Airbnb even introduced a tool, called Create, to make it easier to remix the logo and make it more meaningful for a host's own purposes. "Most brands would send you a cease and desist letter if you tried to recreate their brand," said Airbnb's CEO Brian Chesky when the identity was launched. "We wanted to do the opposite." The Create tool was a little gimmicky, no doubt, but it signaled the way Airbnb saw its community—as a place where you could both belong and be yourself. That's consistent with Marilynn Brewer's behavioral science concept of "optimal distinctiveness," which suggests that the right recipe for building an effective group is making people feel like they are part of it *and* that they can stand out in it.

Along with the new logo, Airbnb retooled its corporate language with a manifesto more like that of an alternative-living community than a Silicon Valley money machine:

> We used to take belonging for granted. Cities used to be villages. Everyone knew each other, and everyone knew they had a place to call home. But after the mechanization and Industrial Revolution of the last century, those feelings of trust and belonging were displaced by mass-produced and impersonal travel experiences. We also stopped trusting each other. And in doing so, we lost something essential about what it means to be a community ... That's why Airbnb is returning us to a place where everyone can feel

they belong . . . At a time when new technologies have made it easier to keep each other at a distance, you're using them to bring people together . . . Belonging is the idea that defines Airbnb, but the way we've represented Airbnb to the world until now hasn't fully captured this.

Contrast this to the classic old power cues in the "About Hilton" section of the Hilton Hotel website:

"Take me to the Hilton." As the most recognized name in the industry, travelers all over the world have been saying "Take me to the Hilton" for almost a century. And because of our innovative approach to products, amenities and service, Hilton continues to be synonymous with hotel across the globe. Hilton Hotels & Resorts remains the stylish, forward thinking global leader of hospitality—and we help make traveling easier with our smart design, innovative restaurant concepts, authentic hospitality and commitment to the global community.

Airbnb's brand voice is built to cultivate a sense of community and participation, and executives are betting that this will be a key source of competitive advantage—because it makes it far less likely that Airbnb hosts or guests will move to the next platform when one emerges. Airbnb now spends millions holding an annual gathering of thousands of its most active hosts, building solidarity and esprit de corps the way a church or Rotary Club might. Going further, it has invested in supporting local groups of hosts as part of a decentralized "home sharing club," supported by Airbnb but led by its most involved members.

Meanwhile, Hilton relies on the value proposition that has sustained it for decades. Its biggest claim is that its brand is "recognized." It sees its advantage in admiration, not activation. Its value proposition offers style, rather than a sense of "belonging." In a new power world, Hilton knows it must tip its hat to a "global community," but that feels very much like an afterthought.

THE OWNERLESS BRAND

When Henry first had the idea for the philanthropic campaign #GivingTuesday, he had some trouble convincing his colleagues at the 92nd Street Y to go along. "But where is our logo?" he was asked. "We must have the 92Y logo between 'Giving' and 'Tuesday.'"

This, Henry tried to explain, missed the point. If 92Y stuck its logo onto #GivingTuesday, no other organization would want to take part. It would be a popular idea in their local ecosystem but could not grow any further. If they really wanted this initiative to spread, Henry argued, 92Y couldn't own it. Instead of 92Y's logo, there was a simple heart between Giving and Tuesday. The mission of the overall campaign had to be valued over the positioning of 92Y as an institution.

Old power brands created by big organizations are typically accompanied by strict rules—use Pantone color 7507 C, use this version of the logo here and that version there, make sure it's reproduced at least this size, lest anyone miss it. They are steeped in a franchise philosophy, not a movement one. Think how McDonald's enforces sameness. It is telling that the very concept of brand is drawn from the practice of literally stamping an indelible symbol of ownership on an asset.

Yet the only reason #GivingTuesday could grow so much is that so many could develop a meaningful stake in it. Chuck Gates, 92Y's art director, initially recoiled in horror when he saw the heart logo he had designed for the initiative being reworked, often by less talented designers, in hundreds of spin-off efforts. Canadians added a maple leaf, breast cancer charities turned it pink, and a testicular cancer charity even flipped the heart upside down.

Today, however, he takes great pride in the many variations of his work, regarding it as one of his "greatest achievements." For him, #GivingTuesday had become much bigger than a logo. It was a coherent but ever-changing new power brand, generously shared among designers—amateur and professional—all over the world. It was ownerless in the sense that anyone could lay claim to it.

Note that this approach does *not* mean that organizations should become anonymous or completely give up on PR. Yet there is a real skill—especially for old power institutions—in what Asha Curran, 92Y's chief innovation officer and leader of the #GivingTuesday project, thinks of as "unbranding," reducing the old power instinct to overwhelm campaigns or ideas with stamps of ownership, and by doing so allowing them to travel more widely.

Note, too, that there is real payoff to the ownerless approach. Giving Tuesday reached a scale and media profile that would have been inconceivable had it been positioned as 92Y's campaign. It repositioned the 144-year-old institution, too, earning it a spot on *Fast Company*'s "Most Innovative Company" list and helping secure a $15 million gift to name its Belfer Center for Innovation & Social Impact.

Step 3: Lower the barrier, flatten the path

In the twentieth century, participating in politics or social change could feel like a test of will. To join a political party, you'd often have to pledge your loyalty and support for a long list of policies, some of which you might not actually agree with. To participate, you'd have to show up to procedural meetings at your local branch where your most exciting moment might be seconding the motion to approve the minutes from the last meeting. And for this privilege, you literally "paid your dues"—up front, and then at some regular interval.

If you were an activist, credibility came from how much you laid on the line—the heroes were the people who put their bodies in physical jeopardy for the cause. In the early environmental movement, you protested logging by chaining yourself to a tree. These weren't easy things to do (especially if you lived far away from the locus of action, or were a not-so-limber grandparent, or just wanted to do your bit without doing anything too extreme). And this meant that many movements tended not to scale beyond small, often more radical groups of activists. Those who did succeed in building a mass base, like the U.S. civil rights movement, drew on a big existing infrastructure, as Dr. Martin Luther King Jr. did with black churches.

But the new power era has promoted a form of activism that can be scaled much more quickly than a twentieth-century movement, and can be initiated by almost anyone, including those without a formal apparatus or an existing bully pulpit. These movements can more easily include wider groups of people, including those who previously had been left on the sidelines or couldn't easily participate.

This dynamic has not just been playing out in the activism space. A macro theme of our age is that participating in almost anything has become easier, whether we are protesting, taking vacations, or even managing our dating lives. The "dating" app Tinder, famously, has reduced the transaction costs associated with finding a date to a series of brutally efficient left or right swipes. To join, you don't even need to create a profile—Tinder can scrape together our existing Facebook profile info and pictures to make one for us so that we can get to judging and being judged right away. What is common among all these things is that we are seeing the barriers to participation lowered and a heightened focus on improving and streamlining user experiences. In a world awash with competing opportunities in which to take part, achieving "frictionlessness"— removing the barriers to action and engagement—has become the necessary art of anyone trying to build a crowd.

The single most important factor behind the early success of GetUp was how easy it was for people (frantic, information-overloaded, but well-intentioned twenty-first-century people) to participate. No member dues, no pledges of allegiance to a platform, no immediate requirement to take to the streets. Rather, GetUp asked people, on joining, to sign a single online petition on just one issue they cared about. GetUp's organizers were relentlessly focused on the how of this—how they could remove all obstacles to signing a petition, *how* long a petition landing page took to load, how much information they should ask for on sign-up (name, email, zip code). This approach horrified some more traditional groups—wouldn't making it so easy to join just attract the weak-willed and the riffraff? Their sign-up forms required much more up-front information, and so by the time people worked their way through them attrition rates were high. GetUp's mantra (which is also that of Silicon Valley) was to make it as easy as possible to get in the door. In doing so, the dynamic between organizer and participant was flipped. The onus was on GetUp to provide its members meaning-ful opportunities to increase their engagement, not on members to prove their undying commitment to the cause.

An unlikely digital wunderkind has used the same basic logic to produce an enormous surge of new power. Indian anti-corruption activist Kisan "Anna" Baburao Hazare is eighty. He doesn't wear a hoodie. For decades he has campaigned for social justice in the Gandhian tradition—nonviolent protest driven by acts of personal sacrifice, such as "fasts unto death" to protest corrupt officials and unjust laws. In other words, for most of his career he has operated in the classic activist tradition—by doing things most of us would never have the courage or initiative to do.

In 2011, Hazare was staging his biggest campaign yet—in sup-port of what became known as the "Jan Lokpal" legislation, a national anti-corruption bill that would strengthen the power of ombudsmen to hold public officials accountable at all levels of gov-ernment, from the prime minister on down. His campaign was striking a chord with ordinary Indians who were fed up with the small and large acts of everyday corruption that chip away at the trust between people and public institutions.

In early April 2011, after the prime minister rejected his demands, Hazare announced he was beginning a fast until the Jan Lokpal bill was passed. Hunger strikes are a very powerful tactic, especially given their legacy in India. They carry with them enormous moral authority and the human drama involved tends to attract media attention, vital oxygen in any campaign against powerful interests. But there's a weakness to hunger strikes, also: they don't give other people anything to do except express support (at least not unless you're willing to do the same). If the job of an activist is to feed the agency of his supporters, Hazare knew he needed to do more.

Hazare began to experiment with new tactics. He asked Indians to send him an SMS if they supported his campaign. Most Indians have a mobile phone, and among the emerging middle class who formed the backbone of Hazare's supporter base mobile penetration was near universal. So Hazare got himself a short code and generated about 80,000 texts from ordinary Indians, a respectable effort.

Then Hazare subtly changed his tactics. All over India and in many other parts of the developing world, people use "missed calls" to communicate with each other. If you're running late for a coffee with a friend, you leave them a missed call. If you're dating someone and just want to let them know you're thinking about them, you leave a missed call (yes, that's right, in some cultures the way to show your lover you care is by hanging up on them). Why do people leave missed calls? Because unlike phoning or texting someone, it's free. It's also effortless.

This small change in tactics made a spectacular difference. When Hazare provided a local number to call and asked Indians to show their support for his campaign against corruption, his numbers went from 80,000 to 35 million.

Thirty-five million missed calls. That's one of the largest single coordinated acts of protest in human history (we've yet to find a bigger one). How did it happen? The ubiquity of the mobile phone is obviously key—ten years earlier, such a campaign would have been impossible. It was also an extraordinary show of strength for India's emerging middle class. And it's a great example of how to build a crowd in a new power way—and central to it was using an existing behavior (the missed call, which everyone was already doing) and by making the barrier to participation low. Taking part was truly frictionless.

Step 4: Move people up the participation scale

You might ask, so what? What do 35 million calls really add up to? Isn't this just clicktivism? Malcolm Gladwell wrote a long missive in *The New Yorker* to this end, arguing that online activism makes it so easy to participate that it all ends up being based on "weak ties," unlike "strong ties" activism where people put their lives on the line and form deep, face-to-face relationships with each other. In one sense, Gladwell is right: commitment matters. As we'll explore in the next chapter, the "super-participants" in any movement or new power community do some of its most important work. But this new capacity to get many more people in the door comes with incredible opportunities for impact, as Hazare discovered.

For Hazare, a mass base of support mattered a great deal. First, those huge numbers delivered tremendous moral legitimacy; these were just too many people for the government to ignore. Even more important, Hazare turned those phone numbers into real, on-the-ground power. Two weeks after making this call to action, Hazare's campaign had the world's largest spreadsheet of phone numbers of supporters. What next?

Hazare's then campaign contacted people on his list to help turn out hundreds of thousands of people to real-world protests in Delhi and other cities. It was an extraordinary display of mobile-to-offline power. And while Hazare's bill didn't pass as is, the government did accept a number of his demands. His 2011 campaign helped to push through sweeping changes to India's laws against corruption. /The Rules, a group focused on empowering citizens in the global south incubated by Purpose, the organization Jeremy leads, then built a piece of technology to help others do what Hazare did: a tool called /Crowdring that allows campaigners in other countries to create their own missed-calls campaigns.

Gladwell's critique of "low-barrier" activism makes little sense if movement builders know how to move people from more superficial to more robust forms of engagement, as Hazare was able to do. This still is baked into what we call the *participation scale*.

On the far left of the scale are the definitive old power behaviors: complying and consuming. Most old power models don't require much more of us than this. File your taxes. Buy these shoes. Renew your subscription. Of course, neither compliance nor consumption is going out of vogue any time soon. Many of our most important social and economic institutions still mostly run on them.

The Participation Scale

OLD POWER BEHAVIORS:		NEW POWER BEHAVIORS:					SHAPING
						PRODUCING	
					FUNDING		
				ADAPTING			
			AFFILIATING				
		SHARING					
	CONSUMING					Creating or Delivering	Shaping or
COMPLYING		Sharing Other People's	Endorsing or Joining a	Remixing Other People's	Affiliating with Money within	Content or Assets within	Protecting the Norms of a
Traditional Compliance	Traditional Comsumption	Content or Ideas	New Power Community	Content or Ideas	a New Power Community	a New Power Community	New Power Community

But if you're trying to build a movement or grow a crowd, you'll need to unlock a series of new power behaviors. You get people in the door via simple, low-barrier asks toward the bottom of the scale—for instance, by asking people to consume and then share content, or by affiliating: in GetUp terms, by signing your first petition. Once you have recruited these new participants, the job is to keep them engaged and to move people up the scale, toward higher-barrier behaviors like adapting or remixing the content of others, crowdfunding a project, creating and uploading their own unique content or assets (we call this producing), or, at the top of the scale, by becoming a "shaper" of the community as a whole, with the capacity to influence the strategy, norms, and culture of a crowd, often without having any kind of formal authority. Think here of the Airbnb super-hosts who set norms for others on the platform, the significant but informal role played by the Black Lives Matter founders, or the most influential volunteer moderators on Reddit, whom we will learn more about in the following chapter.

TED, the conference community, is adept at moving its users up the participation scale. To get people in the door, TED asks us to view (consume) its most compelling TED Talks, drawn from its official conferences and TEDx, its locally organized conferences. It then encourages us to share those talks, and even offers viewers a tracker to help them see how many people they've reached by sharing—a clever way to increase our sense of agency. Moving up the scale, we are asked to join the TED community and affiliate in a variety of ways, for example by nominating someone for the TED Prize. In doing so, we are encouraged to think of ourselves as "TEDsters," a branding device to help create shared identity. TEDsters have an opportunity to adapt TED's original content by joining its army of volunteer translators who interpret the business jargon and pop science that TED is best known for into their own languages. Moving further up, TEDsters are given a range of ways they can fund TED's work, including by attending TED's high-priced official conferences or the much more accessible TEDx events. Finally, TEDsters are able to become producers by organizing their own TEDx conferences or giving their own TEDx talks. There have now been over 20,000 events organized around the world from Madrid to Manchester. The most committed and active TEDx organizers become shapers of that community, helping to steer its direction at global gatherings convened by TED.

Only a tiny fraction of TEDsters move all the way up the participation scale from consuming its content to become TEDx superorganizers, but providing this path makes it possible for TED to move people from content consumers and fans into more valuable roles, like funder or organizer.

Having a structure in place to move people up the participation scale is critical even in the early stages of building a new power community. Perhaps the most inane example of a failure to do this was the social networking app Yo, launched in 2014 by an Israeli software developer as an April Fool's joke with just one incredibly low-barrier feature: you tapped once and the word "Yo" would be automatically sent to a friend in your contact list. Your friend could then "Yo" back, if he or she so chose. Inevitably, the app captured the imagination of the internet. By mid-June the app was the number one social networking app in the Apple app store, and number four overall. The app raised seven figures in venture capital, to the bewilderment of many. Tech blogger Robert Scoble described Yo as the "stupidest, most addictive app I've ever seen in my life." But addictiveness wasn't enough. Yo became a revolving door for users. By September, just a few months after launch, it was down to number 1,277 in the app store. The app's developers used their venture capital money to try to define a path up the participation scale beyond Yo-ing, such as adding location and photos with your Yo and creating groups to make the app more social. But because those steps weren't in place during Yo's initial surge, the opportunity passed to build on that early virality.

Today's most cutting-edge activists understand both how to create frictionless entry points and move people up the participation scale. Black Lives Matter has built the most effective movement for racial justice since the civil rights movement by combining what Gladwell would deride as "weak ties" activism—very simple forms of affiliation and sharing like hashtags that have rapidly spread its message—with more committed actions all the way up the participation scale, from funding like-minded activist groups to "producing" real-world protests and setting up local chapters.

Step 5: Harness the three storms

The American civic leader John W. Gardner believed that "civilization is a drama lived in the minds of a people." His insight helps us in thinking about how new power communities and movements grow. A secret of the most successful movements is that they build up not just in small, steady increments but off the backs of galvanizing moments of drama and urgency that can be hard to predict. The Women's March on Washington the day after Trump's inauguration has become an iconic example of a wave of new power that built in the wake of a historic disappointment—the failure to elect the first woman president, and the election of a man with a long track record of sexist remarks and behavior.

When growing a crowd, good organizers take advantage of such moments. Sometimes, those moments happen to them and their job is to embrace them, even if at first they seem like setbacks, not opportunities. Sometimes, they see a moment out in the world and use it to fuel the movement. And at other times, they create a moment out of thin air. We think of this as harnessing the three storms.

Storm creating

Beatriz Ehlers was a student in the Friedenreich, a municipal school rated as one of the top ten schools in Rio. The Friedenreich was known for serving disadvantaged Cariocas (the term used by the proud citizens of Rio to describe themselves), including many young people with disabilities. Like many community spaces in Rio, the school had been condemned with little warning and marked for imminent demolition to make way for Olympic development, in this case to build a deeply uninspiring parking lot.

Young Bia, as she was known, desperately wanted to save her school. So she used an online tool called Panela de Pressão ("Pressure Cooker") that allows Cariocas to start campaigns on issues that matter to them right from their phones. The insight behind Panela de Pressão was that in order to have any chance of participating in government effectively, citizens needed a way to figure out the right decision-maker to target, then to join with others to put pressure on that person. The tool was a product of Meu Rio, an anti-corruption and civic participation group in Rio, incubated at Jeremy's organization Purpose and the brainchild of a brilliant young woman from Rio, Alessandra Orofino. More than one in ten millennial-aged Cariocas are now members of Meu Rio, and the group is part of a wave of bottom-up protest and civic engagement that is reshaping Brazil's atrophied politics.

When Bia started her campaign, it struck a chord with Meu Rio. How could such a precious school be demolished so cavalierly? they asked. When they saw the early traction that the girl's campaign was getting, they decided to try to turn her fight into what movement builders call an "iconic impact story."

So Alessandra and her team did something dramatic, even theatrical. They set up a website with a twenty-four-hour live video stream of the school, and asked Cariocas to monitor the feed around the clock for signs of bulldozers and other demolition equipment arriving, which was expected any day. Meu Rio then asked people to sign up to become "Citizen Guardians" of the school by registering their mobile phone numbers with the site, and to ask their friends and neighbors to do the same. As soon as the bulldozers arrived, the Citizen Guardians would be urged by text to rush to the school to form a human barrier to protect it, forming a circle of defiance.

The campaign captured the imagination of the city. Thousands of people signed up to protect the school and a huge wave of media coverage followed. Suddenly, some serious heat was directed at the decision-maker with the ultimate power to save the school, the governor of Rio, who had until then been able to ignore the pleas of a few students, teachers, and parents in a disadvantaged area.

Within seventy-two hours of the launch of the live stream, the governor had a rather remarkable change of heart. He decided to cancel the demolition until a new school could be built. And after six more months of persistent campaigning by Meu Rio, the governor wholly capitulated: the Friedenreich would be saved. Bia had won a great victory, but perhaps more important, a signal had been sent to Rio's politicians: the next time you try to demolish a school without consulting the community, people will hold you accountable (and we'll capture every moment on camera). This little victory also triggered a larger debate in Rio about unaccountable Olympic development, one that hadn't fully broken through until Bia's moment crystallized the injustice. And by turning her story into a cultural flash point, Meu Rio had created a storm that built its crowd and advanced its movement.

In 2010, GetUp created its own galvanizing moment in an effort to draw attention to Australia's inhumane treatment of refugees.

At the Australian parliamentary press gallery's annual charity dinner, "dates" with different politicians are auctioned off and usually bought by corporate interest groups seeking face time with elected officials. When then-opposition leader Tony Abbott, an avid surfer and fierce opponent of refugee rights, put a personal surfing lesson up for auction, GetUp decided to hijack this ritual. It asked its members to crowd-fund a bid at the auction. If GetUp won Abbott would be obliged to give a surfing lesson to Riz Wakil, a refugee who had fled Afghanistan in 1999 and endured almost a year in one of Australia's notorious immigrant detention centers. Within three hours GetUp members had chipped in $50,000 and won Wakil a very public date with Abbott. It had found a way to capture the imagination of the media on a vital topic—and give a cheeky win to its members.

Storm chasing

Sometimes a storm is in full swing, and the job of the crowd builder is to chase it down and harness its power. That's how Reddit, the hugely popular platform where people "upvote" their favorite content, really took off.

Five years after its founding in 2005, Reddit was still a bit player among user-driven aggregation sites. Its much bigger sibling was Digg, where tens of millions of monthly visitors busily voted for their favorite content. In 2010, Digg announced "version 4" of the site, a major upheaval that deprioritized user-made content in favor of promoting content from big online publishers like Mashable .com, and introduced "sponsored links," a way to monetize the site. At around the same time, a small subset of Digg users had figured out how to game the system so that the site's article recommendation features shut out content from ordinary users in favor of the group's preferred links, essentially crippling the democratic quality of the site. On August 30, 2010, "quit Digg day" was declared by another set of users, agitated by all of the change, who flooded the site with content from its smaller competitor, Reddit. Reddit saw the storm and chased it brilliantly—Reddit's co-founder Alexis Ohanian wrote an open letter to Digg's co-founder and CEO Kevin Rose, lamenting (with barely concealed glee) that it had strayed from its original mission of "giving power back to the people." Reddit temporarily changed its logo to incorporate the Digg shovel in order to welcome its users, a superb flash of new power branding. Quickly and spectacularly, millions of users migrated from Digg to Reddit. The great Digg exodus killed its traffic numbers and a business once valued at $200 million was eventually sold in a fire sale for $500,000. Reddit as we know it now was born. (In the next chapter, we'll talk about how Reddit then managed—and mismanaged—the crowd it built so quickly.)

The art of turning someone else's crisis into your opportunity is far from new, but the new skill is doing it in an age when the opportunity to mobilize a crowd comes and goes in minutes. The American cookie brand Oreo understood this when a power outage in the stadium caused lights to go out for more than thirty minutes during the 2013 Super Bowl. Oreo struck quickly, tweeting a picture of a lonely but well-lit Oreo and the caption "you can still dunk in the dark." The tweet went viral, "won the Super Bowl," and became the stuff of legends in marketing circles. Often left out of the story is the fact that Oreo had a fifteen-person social media team at the ready during the game for exactly this kind of rapid response.

We don't all have Oreo's resources, but its handiwork reminds us that storm chasing requires infrastructure to do well. Anyone building a crowd should be on the lookout for storms that might galvanize their base, and be prepared to act on them within minutes or hours, when the need is greatest and people's emotional response is at its peak. Many old power organizations will take days just to cobble together a press release. But organizations now need to be set up to move faster, to soak up the energy in a moment and turn that into new supporters. Byzantine bureaucracies requiring multiple sign-offs aren't the right tools for storm chasing.

Storm embracing

Sometimes the best way to build a crowd is to embrace the fact that a storm is at your doorstep and let yourself get swept up in it. As we discussed earlier, the torrent of attacks on GetUp from prominent politicians and the conservative media were key to GetUp's early growth.

Our favorite example of storm embracing comes from an inspired group of Girl Scouts. In 2015, a donor made a $100,000 contribution to the Girl Scouts of Western Washington, targeted for financial aid to poorer scouts—but with a stipulation that the funds not be used to benefit transgender girls. The Girl Scouts felt this would be a grave injustice and essentially said "Bring it on." They returned the funds and launched #forEVERYGirl, a crowdfunding effort to make up the difference in its budget, including aid to young transgender women. The scouts raised $338,000, gained a big new following, and spread a powerful message of inclusivity. One of the backers captured what this campaign meant to her:

> I participated in 15 years of Girl Scouting . . . Twice in five years I've seen you stand up for trans people like me. It makes me cry and brings me such happiness that I could be part of something so . . . amazing. So thank you. So much. For giving love, family and friends to those who are just trying so hard to be true to themselves. For upholding everything I was ever taught Girl Scouts was. You are my heroes.—Quinn, #ForEVERYGirl Supporter

Even big, risk-averse companies are recognizing the value of storm embracing where they might once have tried to shy away from controversy. In 2017, when conservative provocateur Ann Coulter attacked Delta on social media for changing her seat on a plane (featuring an intrusive video of the woman who had been allocated the seat in her stead), the company chose to raise the stakes, responding via Twitter that her actions were "unnecessary and unacceptable." While this alienated some of Coulter's support-ers, the airline's boldness won it rare positive attention. Vox's Mat-thew Yglesias summed up the reaction: "I can't believe Ann Coulter has managed to make normal people sympathize with a Big Three airline."

Whether an organization is creating, chasing, or embracing a storm, some lessons stand out. For one, urgency matters. Research on crowdfunding shows that deadlines are everything—most funds raised in crowdfunding campaigns are in the final seventy-two hours. We've seen many despondent movement builders or crowd-funders discover that people really do come through if you give them a deadline and communicate a sense of urgency.

David vs. Goliath tales are ideal fodder for crowd building, as are crises that threaten the most deeply held values of the commu-nity. Defeats matter, too. In 2013, when Airbnb lost a big legal battle on the rights of hosts who are renters to use the service in New York, it dramatized that moment to rally hosts around the country and explain the long-term challenge it faced. It sees that loss in New York as a critical milestone in its work to mobilize hosts to person-ally lobby in the face of regulatory roadblocks.

Storms can end up as light drizzles, too. And that's OK. Hazare only stumbled onto missed calls after trying short codes; less than 10 percent of GetUp's actions accounted for more than half of its member growth in the early years. A new power community that is growing successfully often has a small, incremental growth rate that ticks along month to month, year to year.

What are the implications of a world in which movements are everywhere, and everyone wants one of their own?

We are witnessing the rise of "WeWashing"—a phrase coined by our friend Lee-Sean Huang to describe the way brands like Pepsi are using the language of the crowd without having any meaningful interest in engaging with it.

As the skills of building a crowd become more widely shared, the public realm is increasingly becoming a kind of Hunger Games for organizations and brands, with each trying desperately to get the storms headed in their direction. In this contest, those who are most rewarded will often be those who stand out: the loudest, the funniest, the most provocative, the most daring. This won't always produce the best social outcomes or the most thoughtful debates. This dynamic can get creepy and problematic quickly: witness the sorry tale of Baltimore's Mike Martin, who built a huge following with his "DaddyOFive" videos on YouTube, off the back of the ever-more-cruel pranks he and his wife, Heather, played on their children. In one video, the parents sprayed invisible ink on their son's bedroom floor then screamed at him for five minutes as he sobbed. In another, the son was encouraged to slap his sister in the face. Mike and Heather ended up losing custody of the children after other YouTubers rallied to protect the kids from abuse. In their belated mea culpa video, Heather explains how their tactics ratcheted up in their quest for an ever bigger crowd: "Once people started watching us . . . we went from something that wasn't so bad . . . and then we just kept going more and more for the shock factor . . . to see what could get more views."

But in this world of constant recruitment, the most effective crowdbuilders will be those who are able to move people up the participation scale, and sustain and nourish a community over the longer term, dealing with the many challenges, compromises, and balancing acts that requires.

That critical capacity is the topic of our next pair of chapters.

WHAT MAKES NEW POWER COMMUNITIES WORK (AND WHY THEY SOMETIMES DON'T)

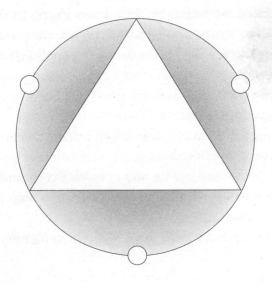

The graphic above isn't some kind of occultist symbol or anarchist cri de coeur. It's a framework that we will unpack in this chapter and the next to help us understand how new power communities work. Many of today's new power platforms are as economically powerful as megacorporations and as populous as many nations. Our daily lives and our livelihoods are wrapped up in them. So it's vital to understand how they work, the roles we play within them, and the impact they are having on society.

In this chapter, we'll explore the roles and dynamics that unfold within these communities—what makes them thrive, what makes them fracture, what makes them fizzle. In the next one, we'll look at what happens when they interact with, and in some cases collide with, the wider world.

We begin our study with the tale of a new power community that fell badly out of sync.

THE STORY OF THE #REDDITREVOLT

"Dear Reddit, you are starting to suck."

The post came from qgyh2, a dedicated Reddit user. Previously, he had been responsible for the highest scoring post in the site's history when his message titled "test post, please ignore" triggered the rebellious instincts of the Reddit community, who cheekily and en masse denied his request.

But now he was fed up.

He ticked off his complaints: Reddit had gotten rid of good staff who cared about the community. It had hired the wrong people. It had started to ignore its users.

For him, Reddit had lost its way, prioritizing corporate interests over all else. "Reddit, when your investors sit in their special cave and count their money they list reddit.com as their main asset here. Reddit.com is not your main asset. Reddit's users and community are your main asset."

He cautioned Reddit to learn from the fate of Digg—an earlier online community—which imploded when its users jumped ship.

And he concluded, simply: "So please stop being a dick and get your act together. Thank you, Q."

On July Fourth weekend in 2015, the ninth most popular website in the United States ground to a halt. Server issues? A malfunctioning cloud? A cyber-attack? None of the above. The cause was user mutiny.

The hugely popular social sharing site Reddit describes itself as the "front page of the internet." Its premise couldn't be more new power: users share links they find elsewhere on the internet and then "upvote" their favorite content (or downvote what they don't like), essentially curating their own website with almost no editorial interference. It is, in many ways, the anti–*New York Times*. More than 21 million votes are cast each day. To put that in context, one week of voting on Reddit chalks up more votes than a U.S. presidential election. It is a huge engine of participation.

The building block of each Reddit post is a piece of content served up by users to delight, outrage, move, or provoke. An image of forty types of pizza. A video of policemen meditating. An interesting question like "what one thing only happens in your country?" If you're not one of Reddit's more than 200 million monthly users, you might not realize that many of the memes, videos, and ideas that drift in and out of your social media stream, and increasingly shape our popular culture, started on Reddit, probably a few days before they hit Facebook or your local TV news. Reddit also has a big impact on politics—research has shown that Reddit was one of the most important online spaces for alt-right Trump supporters to congregate and coordinate as they successfully waged the so-called "meme wars" on social media during the 2016 elections.

The site is much more than a race-to-the-top anarchic free-for-all. In fact, Reddit is very carefully structured. It is divided up into "subreddits," areas for people to discuss and engage around specific topic areas. There are over one million of them, ranging from those you might expect like /r/climatechange or /r/the_donald, for fans of the president, to a remarkably active group devoted to sharing images of "morbidly beautiful things"—/r/morbidlybeautiful. Crucially, the site doesn't have control over those subreddits. They are run by volunteer moderators from among its users. This structure gives Reddit's users and, in particular, its moderators remarkable agency compared to, say, the control Facebook exercises over what people see and engage with. Moderators are in charge of setting direction and shaping discussions, with a wide range of tools and choices placed in their hands. The "Reddit rules," such as they are, prevent illegality or manipulation of the voting system, but other than that, it is up to the moderator. Imagine Reddit as the Wild West of the internet, with the moderator acting as bartender and sheriff.

In addition to the moderators (the "mods," as they are known), the Reddit world is made up of its users, who skew young, male, and geeky, and do the upvoting, linking, and commenting; and its "admins," the Reddit Inc. paid staffers who are charged with making the Reddit community thrive *and* with advancing the interests of Reddit Inc.—goals that, as we'll discuss, are not always compatible with each other, or with those of the community.

One of the most popular subreddits is the "Ask Me Anything" (AMA) thread, where everyone from Barack Obama to Bill Gates takes questions from the Reddit community. As we write this, a ninety-something German woman is starring in an AMA about her experiences as a nurse during World War II—it's the second most popular thread right now on Reddit, coming in just ahead of a discussion about criminal justice reform and a link to a skit about Beyoncé on *Saturday Night Live*.

Victoria Taylor joined Reddit Inc. in 2013 as communications director. She then became director of talent, overseeing over 2,500 AMAs and serving as the company's interface with many of its most important moderators. She was beloved by the "mods," who saw her as more than just an enforcer on behalf of Reddit Inc. but as something more like a friend. "When my wife got pregnant she sent us chocolate-covered strawberries in the mail . . . Just all these little things she would do for everybody," recounts Brian Lynch, a key moderator on the I Am A ___ Ask Me Anything subreddit (IAmA), and a central node in the Reddit volunteer community.

On July 2, 2015, however, Victoria was abruptly fired by Reddit without public explanation. Brian found out when he got a call that holiday weekend saying that there was no admin available to host an IAmA. This was the first he'd heard of the dismissal of someone he saw not only as a friend but as an important ally. For him, and for many other mods, this was an outrage that capped years of deteriorating trust between them and the Reddit platform. They had long suffered broken promises to build features to help them do their jobs better, such as tools to fight spam, and promises of faster response times from administrators. As unpaid, but highly valuable, volunteers, the mods often felt taken for granted, and saw in Victoria qualities that were otherwise absent in their relationship with Reddit Inc. She made them feel like a priority. She recognized their value.

For Brian and his fellow mod Courtnie Swearingen, Victoria's firing was the last straw. So they scrambled the moderator team on a private Slack channel away from the site, where the company couldn't track their conversations. And they decided to deploy one of the most symbolic—and damaging—tools in their artillery. They made their IAmA subreddit "private" in protest, essentially shutting it down. Their idea began to spread. "Other mods said, 'We'll go black in solidarity with you,'" recalls Brian. Soon, the major /r/ science subreddit went dark. Next, /r/AskReddit, the most popular subreddit on the site, shut down. "Then after that it just kind of snowballed where it became the cool thing to do." A rolling blackout of the site's 300+ top subreddits ensued and a full-scale rebellion—what quickly become known as #RedditRevolt, then "AMAgeddon"—was on.

Reddit's traffic had been all but annihilated—by its own moderators.

In the days that followed, the best and worst of Reddit was on display. Reddit's relatively new CEO Ellen Pao quickly apologized, recognizing the ignored grievances of the site's moderators:

> We screwed up. Not just on July 2, but also over the past several years. We haven't communicated well, and we have surprised moderators and the community with big changes. We have apologized and made promises to you, the moderators and the community, over many years, but time and again, we haven't delivered on them. When you've had feedback or requests, we haven't always been responsive. The mods and the community have lost trust in me and in us, the administrators of reddit.

Ironically, Pao's mea culpa was thwarted by the democratic logic of Reddit's own home page: her post was "downvoted" over 3,000 times, making it hard for users to even find it. At the same time her apology was disappearing, a Change.org petition urging Pao's resignation started bubbling upward on the site, soon gathering more than 200,000 signatures. The petition refers to "empty promises to moderators," but it also attacked Pao for something quite different: "Reddit's new age of censorship."

Taylor's firing, it turned out, had brought to a boil a full range of tensions simmering in the Reddit community. In the months prior to the revolt, Pao had been attempting to create new rules to shut down offensive content and hate speech on the site. Predictably, this had clashed with the somewhat outlaw, libertarian ethos of Reddit. Many suspected that her efforts were not driven by a moment of moral awakening but rather were an effort to clean up the site so it could be more easily monetized for advertisers (in one of the great ironies of our era, the parent company of this anarchic online community is Condé Nast).

The speech that Pao was targeting was extreme. She was seeking to ban "revenge porn," a form of online abuse predominantly targeting women, and subreddits like /r/transfag, a poisonous forum mocking and attacking transgender people; /r/shitniggerssay, one of several racist and white supremacist threads on the site; and /r/fatpeoplehate, which had over 5,000 subscribers at the time it was closed down. The ban was framed as an anti-harassment policy, and was directed at moderators who weren't effectively policing their own communities: "We will ban subreddits that allow their communities to use the subreddit as a platform to harass individuals when moderators don't take action."

The users whom Pao sought to silence made up a very small part of the overall community; Reddit co-founder Steve Huffman estimated them as the "toxic 0.2%" of Reddit, polluting the space for everyone else. But far more than 0.2% of Reddit's users were turning against Ellen Pao. The backlash against the ban was not solely about the issues at hand but became a larger statement of protest against the cultural impositions of Reddit on its community. In a space where new power values of openness and freedom were deeply cherished, the old power *diktat*—however well justified—was always going to land hard. But there was something else going on, too. It was also easy to sense gender dynamics at play on a platform whose users were mostly male.

Pao was no low-profile new hire—she was the woman who had just famously sued her former employer, blue-chip venture capitalist firm Kleiner Perkins, for gender-based discrimination. The language of the Change.org petition had more than a hint of misogyny: it cites the lawsuit and refers to Pao as "a manipulative individual who will sue her way to the top." And in the days that followed, Pao was subjected to the very online harassment she had sought to ban on the platform, including death threats from Reddit's users and vicious personal attacks, some of them sexually and racially charged.

By July 10, a week after AMAgeddon had begun, and amid increasing concerns that Reddit users would begin fleeing en masse to other services (the very dynamic that Reddit had benefited from when its rival site Digg suffered a mass desertion in 2010), Pao resigned. A user revolt had toppled the leadership of a major internet platform, and Huffman, the Reddit co-founder, was asked to step back in as CEO to save a new power community in crisis.

SO WHO DOES WHAT? UNDERSTANDING THE STRUCTURE OF A NEW POWER COMMUNITY

How did it come to this? And what can Reddit's travails teach us about how to build and sustain an effective new power community beyond its initial surge?

Any new power community has three key actors—its *participants*, its *super-participants,* and the *owner or stewards* of the platform. Think of these as three corners of a triangle.

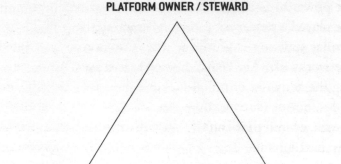

PLATFORM OWNER / STEWARD

SUPER-PARTICIPANTS   PARTICIPANTS

The platform owner or platform steward: The company Reddit Inc. is the platform owner for Reddit. It sets the overarching rules. It owns the brand's IP and pockets the ad revenue. Victoria Taylor worked for Reddit Inc. To take some other examples, Airbnb's platform owner is Airbnb Inc. Facebook's is Facebook Inc. (and, in effect, Mark Zuckerberg himself, who retains effective control of the company). Likewise, Wikipedia, which might feel "ownerless" to its users, is in fact governed and controlled by a board with the power to fundamentally alter its superstructure and rules, something its volunteer editors and users can't do. Platform owners have the ability to control—or at least substantially influence—who is allowed to participate in the platform, its governance and decision-making; how value is distributed; and even whether the platform lives or dies.

Some new power communities don't have any person or entity that meets this definition of owner. Instead, those communities have what we call *platform stewards,* who play recognizable but sometimes informal leadership roles that allow them to channel the energy of the broader community, create rules or norms, and define the structure of a platform. Even the most radically decentralized models, like the virtual currency Bitcoin, have seen such figures emerge. Though anyone is free to take Bitcoin's code, adapt it, and create a new protocol for others to follow, only a handful of people have the power to "commit" code to the Bitcoin code base. These few have played a stewardship role in directing the technology.

A similar function is played by the heads of decentralized terrorist networks like Al-Qaeda—its top leaders are not necessarily directing the activities of individual cells, but they have the ability to set rules, norms (such as they are), and the general direction of the network, even if they can't be described as genuine "owners."

Super-participants: The energy of a new power community is driven by its super-participants—the most active contributors to the platform, and often those who create the core assets that power the platform and create its value. In the case of Reddit, the super-participants are the moderators, people like Brian and Courtnie who shape much of the content and conversation on the platform, and contribute to the development of the community's norms. For Airbnb, the super-participants are the hosts, who offer their properties online. The super-participants for Uber are the drivers. For Wikipedia, the super-participants are the editors, who create and improve its content.

As the most engaged users and those who contribute the most—and who often have the most to lose—super-participants are some of the most influential voices inside a new power community.

Participants: Finally, participants are the people who "take part" in a new power platform, and tend to form the vast majority of its users. For Reddit these are the readers, voters, and sharers of content. For Airbnb it is the guests. For Facebook it is anyone with a profile. For Uber it is the riders. In the terms of the participation scale, participants generally take part by consuming, sharing, adapting, affiliating, and funding, but not by producing assets or playing an organizing role in the way that super-participants do.

The great challenge of the new power community is to weigh the needs of the three different groups, which can easily fall into conflict.

The "Reddit Revolt" was the result of a critical failure of balance in the Reddit community. Reddit Inc. had underinvested in its super-participants for years before the revolt, sowing the seeds for the events of that summer. When Reddit introduced self-moderated groups in 2008, it unleashed a huge surge of agency that became the lifeblood of the platform. Brian recalls why he chose to become a volunteer moderator when the site introduced them: "You got to build a community. You build a policy structure that keeps it thriving, keeps it going. That's why Reddit was successful. You feel like you own it." But in empowering its moderators, Reddit also allowed them to create their own fiefdoms, some of which became bigger than most single-purpose websites, with hundreds of thousands or millions of users in an individual subreddit. These factions had the potential to challenge the platform from within.

Courtnie talks about her identity as a super-participant like this: "Reddit is my community center, it's my YMCA." Moderators like Courtnie or Brian weren't demanding money or credit, but they wanted to be supplied with the tools to do their job better, and to be treated with respect by the administration. They asked for simple things like updates to their "modmail," the internal message tool that moderators use to manage their groups. For years, those requests went largely unheeded by the platform, and the relationship between Reddit Inc. and its super-participants gradually deteriorated.

The tension between Reddit Inc. and the moderators was the backdrop for Pao's effort to ban the most offensive behaviors from the site. And the #RedditRevolt brought it all to a head. The platform was battling with its users and its moderators. The moderators had locked out users. Users were at odds with each other. It was quite the mess.

The Reddit story is a dramatic version of the tensions that new power communities manage every day. The needs of platform owners, super-participants, and participants are often not aligned, and sometimes clash directly. Think of Twitter, which has been challenged because its super-participants (the influential super-users who dominate the platform) love its quirky functionality and culture, while those same qualities prevent growth among the vastly larger market of everyday participants, many of whom find Twitter noisy, confusing, and nasty.

To dig into these dynamics a bit more deeply, let's turn to the sharply contrasting ways that Uber and Lyft—two ridesharing apps with very similar businesses—are managing their new power communities. This juxtaposition tells us a lot about the connections among platforms, super-participants, and participants, and the factors that can bring them closer together, or drive them farther apart.

ORGANIZING PICKETS VS. ORGANIZING PICNICS: THE BIG DIFFERENCE BETWEEN UBER AND LYFT

The battle of Uber vs. Lyft has become the Coke vs. Pepsi of the new power economy. The two companies are both chasing the same drivers and riders. They live in fierce and unfriendly competition, with Uber well ahead, having scaled much faster and expanded globally, leading to a valuation over ten times that of Lyft, but with Lyft posing a real threat in some of Uber's biggest markets.

The functionality of the two platforms is very similar. An Uber user feels thoroughly at home with the Lyft app and vice versa. But from the beginning, Uber and Lyft have positioned themselves very differently. Uber launched as "everyone's private driver"—the pitch being that you, too, could slink into the back of a badass shiny black ride. Lyft came to life as "your friend with a car," with a giant pink mustache amiably perched on the grille, riders hopping in the front seat and fist-bumping the driver a hello.

Over time, Lyft has mostly ditched the mustaches and the fist-bumping, but still positions itself as trying to get closer to its drivers, and its riders. Uber is defined by its remoteness—with a cutthroat "bro" culture that has created a toxic relationship with its key constituents and led to the downfall of founder and CEO Travis Kalanick. A typical comment from Kalanick, anticipating driverless cars, summed up the company's attitude, casting its drivers as little more than a cost center: "The reason Uber could be expensive is because you're not just paying for the car—you're paying for the other dude in the car." Yet the Uber story goes much deeper than the personal failings of Kalanick. The culture he enabled and represented came to define the company.

To see these different approaches at work, let's consider the ways the two platforms went about implementing rate cuts at the start of 2016, a decision that gave riders cheaper fares but squeezed drivers hard.

How to cut fares and influence drivers

It all began when Uber suddenly announced it was cutting its rates in eighty cities across the United States, anywhere from around 10 to over 20 percent. This was welcome news for riders, but *not* for drivers.

As Harry Campbell—who runs the popular blog *The Rideshare Guy*—observed: "In true Uber fashion, they made the announcement after business hours on a Friday and didn't even e-mail drivers about it." Uber knew best, so Uber set new rules. Its argument was that the cut was a critical measure to drive demand during the slow winter months. It claimed that, statistically speaking, the driver would earn more owing to increased demand. Of course, this logic was not immediately attractive to the driver who used to receive $43.67 to get to LaGuardia Airport from Manhattan and would now receive $37.12.

Uber was taking a classic old power approach: a decision by fiat, handed down to drivers with no warning or consultation, delivered with a seminar in how they might better understand their own best interests. Across the country, reactions from drivers included strikes, protests, and threats of boycotts.

Lyft, weighing a response to Uber's price cut, had a decision to make. When Uber had made similar cuts in the past, Lyft had not, but it had cost them business. For all the fist-bumps in the world, price is the name of the game. As CEO John Zimmer wrote in a message to drivers: "Even with [our] better service, passengers first choose the more affordable option." So Lyft decided to cut fares, too, but looked to their driver community to help guide the process. Bringing together a wide range of drivers in San Francisco, Lyft was able to generate a series of initiatives that would mitigate the cost cutting. One idea was to increase the referral fee for drivers securing new passengers. Another was free car washes. A third was to encourage regular meet-ups for drivers, funded by Lyft, because "the bonds you create with each other strengthen our community, too." At the same time that Uber's drivers were turning against the platform to organize pickets, Lyft was helping its drivers organize picnics.

Every one of the new initiatives was proudly displayed on the Lyft blog, with a driver credited for each idea, alongside testimonials from Lyft *riders,* celebrating what great human beings Lyft drivers are. As @rounditrosie wrote, "I love @Lyft drivers because they are artists, bakers, med students, retirees and the coolest people in LA. #ThankYourLyftDriver." While some of this might feel like sugarcoating on a bitter pill, it worked because Lyft's instincts so consistently demonstrated new power values. The platform owners were transparent and open to super-participants about the financial challenge they faced. They looked to their network—not simply their senior managers—to find solutions. They created structures to get all three corners of the new power triangle allied in facing the challenge, with incentives for their drivers and the hashtag campaign for their passengers.

Inductions and inducements

"Rideshare Guy" Harry Campbell has been a driver for both companies. He explained to us that the stark difference in culture between Uber and Lyft plays out broadly in how they manage their drivers.

For both firms, the ease of signing up as a driver is touted throughout their networks. (Compare the promise of "signing up takes less than four minutes" with the two years of deep study of the "knowledge" needed to become a London taxi driver—a powerful reminder of how our notions of expertise are changing in a new power world.) Yet for all the similarity, the firms take a different approach to induction. As Campbell tells us, "As a new Lyft driver, you actually have to meet with a mentor, so that's another, more experienced driver, and they'll walk you through the ropes, do a spot check on your car, make sure that you'd basically be a good Lyft driver, and take you on a quick test drive around the block." For Uber there is no such procedure for bringing new drivers on board. "It's all done over email communications, so you can imagine that a lot of new Uber drivers really feel like they're out there on their own. They have no interaction with other drivers, they have no interaction with passengers, they have no interaction with a physical person at the company themselves. They feel like they have to figure these things out for themselves. They don't have coworkers that they can talk to, they don't even have someone at the water cooler, because their car is their office. I think that's the sense that you get, right from the onset, that Lyft does care."

Campbell explains that Uber seems at pains to keep itself at a *distance* from the lived experience of their drivers: "Uber actually had a policy where they don't allow their corporate employees to be Uber drivers, where Lyft is almost the opposite. They highly encourage their employees to be drivers."

Lyft tries to show it cares, too, in how it offers drivers incentives. Lyft has always offered riders the opportunity to tip drivers; Uber only introduced this feature in 2017 under pressure from drivers and besieged by crisis. Lyft also takes a different approach to rewarding their most committed drivers, operating a sliding scale that reduces Lyft's commission based on how many hours a driver works. The most dedicated, who chalk up fifty hours a week, "basically get your entire commission back." Not so for Uber.

These differences have a real impact on drivers—who, according to Campbell, largely prefer to drive for Lyft—as well as an interesting effect on the rider-driver relationship. "One of the things that Uber drivers complain about, jokingly, amongst drivers, is that when you get to pick up an Uber passenger, they make you wait forever . . . It's kind of like going off the slogans. 'Would you make your private driver wait?' 'Yeah. Probably.' 'Would you make your friend with a car wait?' 'Maybe not.'"

The power of "platform culture"

The sum of the relationships between platform, super-participants, and participants makes up what sharing-economy guru Arun Sundararajan has called "platform culture," those "shared norms, values and capabilities among the providers." He sees this as the analog of a corporate culture, but "without the directive authority or co-located social systems that traditional firms can take advantage of to manage their employees."

The Lyft story demonstrates some of the critical tactics and philosophies of creating a positive platform culture. It enlists its network participants as allies, or at least signals in this direction. It approaches the world with a human face, not just an algorithmic logic. It is working out ways to align pay and other incentives with a sense of decency and fairness. It invests in getting close to those in its network to better understand their challenges and realities.

Most likely, the duel between Lyft and Uber will be settled by the brutal logic of price and convenience. If it takes Uber three minutes to match driver and rider, and it takes Lyft eight, then the rider looking to get home, and the driver looking to make bank, aren't going to be easily swayed by touchy-feely matters of culture and inclusion. But if and when that gap closes, things might start looking very different.

There is big opportunity, too, if either platform starts to do more to truly benefit its drivers. In an effort to shed its Evil Empire image, Uber is exploring a range of initiatives to sweeten the deal for drivers, including offering them equity in the company.

Lyft co-founder John Zimmer is warmly anticipating the fight ahead. If Lyft can begin to match Uber on product, as they are doing in major cities like San Francisco, where they hold almost half the market, then culture becomes a critical differentiator. His biggest advantage may be Lyft's drivers. In a 2017 survey conducted by Campbell, more than 75 percent of Lyft drivers said they were satisfied with their experience. Less than half of Uber drivers could say the same.

INSIDE THE TRIANGLE: DESIGN CHOICES
FOR NEW POWER COMMUNITIES

Uber vs. Lyft is a story of design choices. How much agency and value should go to participants? How to recognize and reward super-participants, and how to create feedback loops that keep them coming back?

Anyone wanting to understand the dynamics of a new power community—or design one for themselves—needs to wrestle with these key questions.

Who gets rewarded—and who "pays"?

Incentives and rewards in new power platforms shape whether—and how—participation occurs. Uber and Lyft attract riders with low prices and fast pickups. They attract drivers with flow of demand and surge pricing. Yet these two dynamics aren't always in sync. When it rains, say, the market shifts, so the sodden father of two now pays three times the normal rates to get his whiny kids home (and hates the platform every mile of the way), while the delighted and dry driver is now making a morning's take in a couple of hours. The platform must decide how to balance these competing incentives.

To take another example, YouTube's Partner Program shares 55 percent of the ad revenue coming from a video with its creators. This has made a very small number of young video makers quite rich, and created incentives and loyalty (and dreams of stardom) for a longer tail of super-participants on the platform. And Etsy's long-standing commitment to never take more than 3.5 percent of the revenues raised by their sellers is a brand promise at the heart of managing and engaging their super-participants.

It can be hard to find the right rewards to drive a new power community. Sometimes a financial incentive gets the right result.

Sometimes the sense of being part of something important will deliver more than cash. Often you need the right balance of both. Yelp, for example, initially tried to pay reviewers and found it created a transactional dynamic and weakened loyalty, so they adopted a volunteer community strategy instead.

Meetup.com had the opposite experience. Their network allows anyone, anywhere to set up meetings with others in their neighborhood in person, to do anything from practice their French to talk about living with diabetes. When Meetup started, it was free for anyone to start a group, and thousands of events were posted. But what Scott Heiferman, Meetup's founder, quickly found was that the organizers of these free groups were less likely to take their obligations seriously—and sometimes wouldn't even show up to the events they were hosting—if they didn't make a financial commitment to being a host. So now Meetup's primary business model is a modest ($10–$15/month) charge to their super-participants to manage a group. It has been the making of the platform.

Whether money is in the mix or not, the critical task here is to create incentives that not only don't treat participants like commodities, but also reinforce community norms. The first breakout crowdfunding platform, Kickstarter, figured this out early. Those running campaigns are encouraged to reward "backers" in creative, generous ways (like inside access to the campaign's creators) that don't map directly to the monetary value of their contributions. As Kickstarter's founders remind their community: "Kickstarter is not a store."

Who gets recognized and who has status?

Recognition systems that validate the work of participants or super-participants can be hugely motivational. Think about the power of the "validated" Twitter user with a big tick next to their name. Or take eBay's "power-seller" program, which confers both tangible commercial benefits and a somewhat intangible—yet significant—community status.

Recognition sometimes takes the form of seemingly banal and meaningless online monikers and status bumps (like Snapchat's much vaunted trophy case), but even these can be highly effective at driving repeat engagement. On Reddit, recognition is conferred using "karma points," displayed on every user's profile based on how useful the community has found their links and comments. For the outsider, these points could seem meaningless, but for many Reddit users they are a big part of why they feel valued and keep coming back. Well-designed, intangible reward systems can confer huge meaning for the individuals who take part in them.

The trick is to build status in a way that feels consistent with the ethos of the platform; Uber's VIP program for its most loyal riders feels consistent with that brand's elitism, but for some platforms creating a virtual caste system might feel inegalitarian. This is especially true for the new crop of digitally-driven social movements, where the principle of inclusion and equal voice would clash with efforts to put the most active or valued participants on a pedestal.

What creates stickiness?

At the heart of every successful new power platform is a great feedback loop. We are all now hooked on them. From multiplayer games to online petition sites, well-designed feedback mechanisms that track our progress (and that of our friends) and create a drip, drip, drip of dopamine-fueled satisfaction are increasingly central to most modern consumer technology experiences. Think of how the likes or comments on people's Instagram posts keep them coming back. And these aren't (just) ego boosts. Done right, they can make people feel connected to one another and tied to something bigger than themselves.

These feedback loops can become more compelling when there is a time factor or collective goal in play. Think of the urgency of the crowdfunding experience, where you watch your $100 contribution play a role in hitting a $5,000 goal. Or think of the Nike running app, where we can see our own progress and compare ourselves to friends, in a potent mix of collaboration and competition.

How do you establish trust?

Many new power networks simply wouldn't function without the *reputation systems* that manage risk and incentivize collaborative behavior within them. Ten years ago if you had described to someone the basic model for Airbnb—that you'd rent your home, with all the furniture in it, to a complete stranger without meeting him or her—he'd have chuckled heartily. In fact, the celebrated venture capital firm Union Square Ventures famously passed on Airbnb because its partners were skeptical people could ever trust one another enough to do this. But systems that establish reputation and build trust have enabled all kinds of seemingly far-fetched new power models to come to life.

By allowing guests and hosts to rate and provide detailed reviews about one another, and by assuming certain risks as the platform owner (what you might think of as "centralized" trust functions, like payment security and identity verification), platforms like Airbnb have been made possible. Exactly how these systems are designed, however, shapes how effective they are. Airbnb realized that because of fear of retaliation and social awkwardness reviewers were leaving each other much less honest reviews, so in 2015 they changed how reviews worked to have them released simultaneously, so a host's review of a guest wouldn't be shaped by a negative or positive guest review, and vice versa. Researcher Tom Slee has found that, before this change, on Airbnb the vast majority of ratings were 4.5 or 5: "When we rate each other, ratings become more a courtesy than a judgment. Just as restaurant tips only weakly correspond to the quality of service, so a rating of 4.5 or 5 is more a way of politely concluding an exchange than it is of assessing the behavior of a host or guest." This is a trend found on other new power platforms, and you'll recognize it in that interaction you have with your Uber driver when he tells you he's giving you five stars as you get out of the car, in hopes that you'll feel a moral obligation to reciprocate. *feedback system 2*

There's another important thing to keep in mind about the role of trust and reputation systems in new power models: how they can reinforce existing bias or injustices. Ben Edelman and Michael Luca of the Harvard Business School found that "nonblack hosts are able to charge approximately 12 percent more than black hosts" for comparable listings, controlling for ratings and other factors. Further research by Edelman, Luca, and Dan Svirsky showed rampant discrimination against black guests. The very system Airbnb had designed to promote trust—using the real names and faces of guests and hosts—allowed implicit bias (and plain old racism) to cause explicit harm. After a significant public backlash, with black users sharing their experiences alongside the hashtag #Airbnb WhileBlack, in 2016 the company announced a series of measures to fight this kind of discrimination. But notably, they resisted calls to anonymize names and profile pictures because "profile photos are essential to Airbnb's overall mission of building a community."

Who calls the shots?

When members of the Occupy movement in Zuccotti Park made decisions, they gathered for what they called the "general assembly." Decisions required almost complete consensus among hundreds or even thousands of people via an elaborate system of hand gestures: waggling fingers for agreement, limp wrists to the floor for disagreement, crossed fists for an outright veto on a proposal. At the general assembly, everything was designed to create a feeling of radical inclusiveness and a sense that every voice was equally valued, not just the loudest or most powerful ones. The "people's mic" was a tactic used in which the crowd at a general assembly would repeat what the speaker was saying in short phrases so everyone could hear. Initially created as a workaround because the police had cracked down on the use of megaphones, it came to represent Occupy's ethos of participation and collective action. As social movement theorist Craig Calhoun put it, "The human megaphone evoked the decentralized, popular nature of the occupation; it made the group a demonstration of participatory democracy."

Occupy took an extreme stance on who calls the shots: namely, everyone. And it ultimately cost them. Andrew Cornell recounts a sympathetic observer of Occupy LA explaining the loss of momentum there: "Three weeks into the occupation, the group was spending more time discussing its own process than anything else."

Many new power communities—especially corporate ones—can tend toward the other extreme, where the platform owners really call the shots, delegating only trivial choices to the masses. Facebook Inc. makes all the big governance decisions, with the average members just able to tinker with their privacy settings and choose from a range of emojis to demonstrate their mood.

To show how these vital design choices can play out, we'll finish this chapter with the story of a movement that, in its formative years, got most of these questions *right*: building a passionate community and a well-balanced triangle, structured with the right incentives and feedback mechanisms.

THE PERFECT TRIANGLE OF INVISIBLE CHILDREN

In a way, it was all Robert Redford's fault.

The original plan was simple, sincere, and naive. Return from African adventure in 2003. Make a truly compelling documentary. Enter the Sundance Festival. Become documentary darlings. Meet Redford. Maybe have him introduce them to Clooney. Achieve global superstardom. See the world unite against the savagery of Ugandan warlord Joseph Kony. Bring him down.

But Sundance said no. And the three young American founders of Invisible Children, Jason Russell, Bobby Bailey, and Laren

Poole, were stuck. They had a story they felt the world needed to hear. Lives depended on it. Yet they had no distribution. There was no Netflix original programming back then. YouTube was an infant. So, through lack of options as much as anything else, they set out to build their own model. As Ben Keesey, Invisible Children's former CEO, told us, "That was our impetus for giving the finger to the system and saying, we're going to build our own distribution, we're going to go right to the audience."

Their first movie, *Invisible Children: Rough Cut,* the one that Redford passed on, is the origin story of their movement. It chronicled their trip, with three hip young white men, fresh-faced and surfer-preppy, heading to Africa "in search of our story," and finding it as they discovered the atrocities of Joseph Kony, who forcibly enlisted children into his terrorist militia.

Many years later, the group would face criticism for positioning themselves as saviors and heroes, but in this first film they come across as well-meaning, culturally naive doofuses. They went to Sudan and got bored. They blew up a termite's nest for kicks. They killed a snake with a hatchet. They threw up a lot. It was more *Hangover Part II* than "I Have a Dream."

It is not hard to see why Sundance passed. But it is also not difficult to see why many kids would become so inspired. The movie told an underreported story of a monster on the other side of the world. It showed children sleeping en masse at bus stops to find protection from him in numbers. It was compelling and relatable. For young audiences, it was a story of people like them discovering a cause, and then asking other people like them to help.

The role of the "Roadie": How super-participants built the Invisible Children network

To get the word out, Invisible Children started to organize small screenings in churches and on campuses. Think of these as mini-Sundances, where communities of people would come together around the movie to engage with it, discuss it, and—crucially—act on it, offering their voice, funds, and support. The screenings could not offer a huge blast of publicity, but they offered a different kind of value: deep connections to local audiences.

For the next eight years, Invisible Children shaped and perfected this local model. They would typically have two "tours" a year, where they would debut a new movie or highlight a key campaign. In total, they launched 16 tours, leading to over 13,000 screenings that reached an extraordinary 5 million people.

Each of these mini-Sundances relied on their own Redford-like figures. Called "Roadies," these were young people who would pledge six months of their lives to volunteering to organize campaigns and screenings. The recruiting video Invisible Children used to attract Roadies highlights why the job was so enticing. Telegenic former Roadies guaranteed that "you're part of something huge" and you'll "blow their minds," recalling "four-to-a-bed" sleeping arrangements and even quoting Henry V's St. Crispin's Day speech. It was a promise of agency and meaning, dressed up with a henna tattoo and a wink.

The Roadies were earnestly trained, taught the history of the conflict in Uganda, given lessons from the civil rights movement, organized in teams of four, and paired with a survivor, often a former child soldier. By the time they finally set out on the road, they had already invested significant time and energy in the events they were planning, and so had their local organizers, whose floors they would often sleep on.

All this activity was tracked centrally via a Salesforce database so "Roadie teams could see their donations and sales, statistics, compared to other teams, all this internal competition. It was fun," says Keesey. Invisible Children had built smart and transparent recognition systems to help drive the ideal Roadie behavior.

This is a valuable study of how to engage and develop a super-participant. Roadies were highly incentivized and bound together in tight-knit groups by a strong culture. Their job was to build a movement around each local moment, driving engagement, online buzz, and participation. They were invested in each small success, which added up to something bigger than each individual screening.

Once the Roadie team left town, Invisible Children did not leave people's lives. It cultivated the network of local organizers who had held screenings and used this base of people to organize big national moments like 2009's "Rescue" event. The Rescue, held on the heels of the latest film release, was designed to unleash the creativity of young supporters activated by the screenings. As Zach Barrows, Invisible Children's movement director, explained to us: "In order for your city to be rescued, you had to have the media show up to cover your event and you had to have some type of mogul, international or from your community, come out and recognize that this [the kidnapping of children] is in fact happening." This led to 80,000 kids turning out in 100 cities, often accompanied by media-grabbing influencers, to host physical events that were filmed and tweeted to reach many multiples more. To achieve this reach, it helped that many of Invisible Children's early "connected connectors" were young evangelical Christians, a highly motivated and closely networked group.

All these activations had extraordinary effects on the Invisible Children community. The researcher Beth Karlin worked with the group to try to better understand their network, surveying over 2,000 participants, from Roadies to attendees at screenings. Her findings showed astonishing levels of commitment. Over 90 percent had talked to friends/family about Invisible Children, 78 percent had purchased an Invisible Children item, 75 percent had made a donation, and 42 percent were part of an Invisible Children club or group. The Roadies and organizers had done a remarkable job of cultivating engaged participants.

Campus by campus, tour by tour, week after week, Invisible Children had mobilized a nation of young people around an obscure problem on the other side of the world. As Zach Barrows says: "It was like—you thought of Invisible Children and your first thought was . . . I know those people, those are my friends. It was built on personal connections."

These connections had come together in a perfect triangle.

Invisible Children had found the right ways to reward, engage, and affirm all their constituents.

Yet, eight years in, for all this symmetry and energy, they were facing a fundamental problem.

Joseph Kony was still at large.

WHAT MAKES NEW POWER COMMUNITIES WORK (AND HOW THEY FUSE WITH THE WIDER WORLD)

"The next 27 minutes are an experiment . . ."

Invisible Children was frustrated. It had built a reputation for innovative activism. It was beloved by an army of young people. But for its young founders, the mission would not be complete until Joseph Kony was brought to justice.

So, somewhat in desperation, they decided to go in a very new direction.

What they needed was a massive burst of fame and publicity. They needed Kony's name *everywhere*. This, the founders believed, might finally help finish the story they had begun almost a decade ago. Their highly charged but relatively localized campus and school projects just hadn't got the word out widely enough.

They decided to put all their efforts and resources behind a new film that they would distribute differently. Instead of using their tried-and-tested Roadie model to launch it, with each local organizer producing events around the content to introduce it to their community, they would release this film direct to market. Their goal was to create a viral sensation. They called this film *Kony 2012*.

They succeeded beyond their wildest dreams. Within a week of its March 5 launch, 100 million people had seen the video (which, at almost thirty minutes long, defied expectations of what people would watch on a phone or laptop). It was declared "the most viral video in history." Surveys suggested that more than half of young American adults had heard of it.

Kony 2012 was ambitious, inspiring, and dramatic, hitting every emotional note. It was narrated by the magnetic leader of Invisible Children, Jason Russell, who was featured heavily in the film, even turning the camera on his very cute young son, who joined him in promising to "stop Kony." The production values were gorgeous; the cinematic storytelling was worthy of Russell's film school training.

The call to action accompanying the video was smart and, at the time, a real innovation. The founders set up a website making it easy to tweet at "20 culture makers and 12 policymakers to use their power for good." They asked the public to call on people like Mark Zuckerberg, Justin Bieber, Bill O'Reilly, and Oprah to help "make Kony famous."

The video was seeded and propelled by the existing Invisible Children community, and especially its fervent teenage and mainly female fan base. Its hard work over eight years had seemed to pay off; when it mattered, the fans had shown up to promote the cause. As Gilad Lotan of SocialFlow noted, "This movement did not emerge from the big cities, but rather small-medium sized cities across the United States." But from that base of "connected connectors" went into the hands of countless celebrities. Within hours of its exploding, people like Oprah were responding (a little defensively), "Thanks tweeps for sending me info about ending #LRAviolence. I am aware. Have supported with $'s and voice and will not stop.#KONY2012." Major media then jumped on board and started to cover the surge of interest, further spreading word of the movie. None other than Bono declared that the suddenly famous Jason Russell deserved an Oscar. Kony's name was everywhere.

But then something went very wrong. Invisible Children's website crashed under the pressure. So did Jason Russell, who within days of the release of the film suffered a psychotic episode that left him pacing naked in the streets near his home, which was—of course—filmed on someone's phone. That went viral in its own right.

These breakdowns precipitated a huge backlash. An internet

mob took cruel and voyeuristic delight in Russell's mental health crisis. And the initial praise of people like Bono was drowned out by a universe of critics and skeptics who now turned against Invisible Children. There was such volume that major media outlets started to host blogs simply to capture all of the various concerns that were stacking up.

Invisible Children was accused of slacktivism, white-savior complex, impropriety, dodgy dealing, stealth evangelizing, colonialism, paternalism, and being a front for the CIA. Its leaders were attacked for being dangerously naive, stupid, reckless, and insufferably condescending. Influential African voices criticized its work as wrongheaded: chasing the wrong problem at the wrong time. Longstanding NGOs working in the space, some of whom felt their role was being threatened or usurped, started to weigh in. Those initially supportive celebrities moved on to other things. A university student started a very popular Tumblr account, Visible Children, that did a lot to shape a compelling counternarrative, questioning Invisible Children's finances and their strategy.

Pushing back, the Invisible Children team desperately tried to tell their story. But their website was down, their leader was in the hospital, and their nineteen-year-old PR intern, Monica, had no reasonable hope of triaging the 4,000 email requests from the press. For all their years of effort, they were seen only through two narrow filters: the troubling simplicity of their film and the troubled psychology of their founder.

Kony 2012, like their previous films, was designed as the prelude to a coming event, "Cover the Night," where young people would spend the night covering cities with Kony posters. Then, the theory went, the world would wake up and take notice. But it flopped badly. Twenty-one thousand had signed up for the event in Vancouver; just seventeen showed up. The Brisbane event reported just fifty people.

Far from taking Invisible Children to the next level, *Kony 2012* struck it a fatal blow. Although the core team kept on fighting, and even notched some policy wins, the group had been irreparably damaged. Just two and a half years after the film debuted, Invisible Children announced that its U.S. operations would dramatically shrink, bruised by all the bad press and negativity, but doomed by the fact that its "core model of hosting thousands of screenings at high schools and colleges dried up."

The critical mistake for Invisible Children was that they achieved what Robert Redford had denied the founders right back at the start. They had finally released a blockbuster.

What went so wrong for Invisible Children

The pre-*Kony 2012* version of Invisible Children showcases a triangle beautifully aligned among the leadership; its super-participants, the Roadies and organizers; and the participants, the millions of people who attended their events and tweeted, viewed, supported, and shared their work. Their movies were just one element of a richer community experience, a great story that passed peer to peer like an epic poem, strengthened by each telling.

Kony 2012 bypassed that model. And once Invisible Children moved outside the comfort of its triangle, it lost control of both the

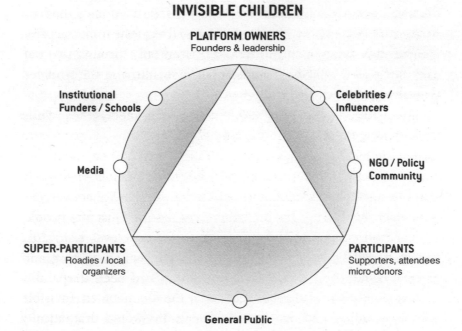

INVISIBLE CHILDREN

PLATFORM OWNERS
Founders & leadership

Institutional
Funders / Schools

Celebrities /
Influencers

Media

NGO / Policy
Community

SUPER-PARTICIPANTS
Roadies / local
organizers

PARTICIPANTS
Supporters, attendees
micro-donors

General Public

model and the message. If we map out its neighborhood post-*Kony*, we can see the triangle surrounded by a circle of interested parties it had failed to cultivate and/or quickly fell into conflict with.

Invisible Children's close-knit community had always seen its founders as lovable rogues, dedicated to humanitarian issues, but the wider world had no such connection. As Zach Barrows remembers, "When *Kony 2012* hit, suddenly millions of people knew who we were, and when their experience with us was just about the movie and they didn't understand the people or the organization behind it, that loyalty factor was definitely gone."

The huge success of *Kony 2012* ignited the interest—and perhaps envy—of a range of slumbering outside stakeholders, all of whom began to stir. As balanced as Invisible Children's community had been, it had also been fairly isolated from key influencers like academics and bloggers who worked on the issue. It had flown under the radar of the major media.

The film also shifted focus away from the community and onto Jason Russell as a charismatic leader figure. He aspired, as a strategy, to connect directly with the likes of Bono and Buffett, enlisting his base to help him do so. But where the old model of Invisible Children made thousands of people feel like leaders, this approach made it all about Jason. His breakdown was thus the organization's. (For another example of this dynamic, think of Lance Armstrong and his now-diminished cancer charity Livestrong.) The superparticipants, who had invested so much in the work of Invisible Children, were left feeling conflicted and disempowered. They didn't have a clear role in this production.

UBER'S VICIOUS CIRCLE

The *Kony 2012* story offers a dramatic illustration of how the circle of actors around a new power community can impact its fate.

We see similar challenges playing out on a much larger scale with some well-known new power models. In the previous chapter, we saw how out of whack Uber's triangle has become. In 2017, as those tensions started to reverberate around its wider circle, they cost founder and CEO Travis Kalanick his job.

Over the years, Uber had almost seemed to delight in picking fights with others in its circle. In 2014, Travis Kalanick explained Uber's worldview this way: "We're in a political campaign, and the candidate is Uber and the opponent is an asshole named Taxi." (It is fair to say, given Kalanick's own personal brand, that Taxi might see its opponent in similar terms.) Initially, Uber could rely on the support and energy of its drivers. But, as *Fast Company*'s Sarah Kessler observed, taxi drivers and Uber drivers began to recognize their common interests in the fight for better pay and conditions. Uber's sudden rate cut in 2016 led taxi drivers and Uber drivers in the United States to form joint protests to avoid a "race to the bottom." Kessler quotes Abe Husein, a former Uber driver who led strikes in Kansas City and elsewhere: "When I first started driving, taxis hated Uber drivers, and Uber drivers hated taxis . . . Now everybody hates Uber. Times have changed."

Uber's move-fast-and-break-shit government relations strategy and its growing notoriety have also made it a hugely popular target for regulators. In 2016 it was fighting more than seventy federal lawsuits in the United States, having settled another sixty. In London in 2017, Uber had its application to renew its operating license rejected.

These tensions were behind the #deletcuber movement that caught fire in early 2017. It began when Uber was accused of trying to profit from rides from JFK Airport in the melee surrounding a protest of President Trump's ban on Muslim immigrants and refugees. This was compounded by news that Travis Kalanick was heading to the White House to be part of Trump's Economic Advisory Council. The campaign went wildly viral, in part because people were already primed to despise and distrust Uber. Not only was Kalanick pressured into resigning from the council, but over 200,000 users deleted their apps. Weeks later, another surge of deletions occurred after a former Uber corporate employee, Susan Fowler, blew the lid off the company's culture of sexual harassment on her blog. Uber was reduced to sending messages to users who were trying to delete the app, pleading with them and saying it was "deeply hurting." The media had a field day with these stories, compounded by more internal strife and the leak of a video of Kalanick arguing with one of his own drivers. Kara Swisher and her popular technology blog *Recode* were a critical voice (in both senses), shifting opinion against Travis and Uber.

Kalanick had raised more venture capital than any start-up ever had, and in the end it was five of Uber's big investors who put the dagger to Kalanick. By the time they intervened, we were witness to a horribly mismanaged triangle and circle. Media, regulators, competitors, staff, drivers, riders, and popular opinion had all turned against Travis. A new power leader who has no support within his own triangle has little chance of surviving a hostile encirclement.

Uber post-Travis is working hard to put a fresh face on its company, but there is likely to be more conflict ahead. Uber's challenges are not just cultural, but structural. In 2016, Uber lost $2.8 billion. These losses are driven by the ride subsidies it must offer all over the world as it pushes for market dominance. Uber will have to increasingly squeeze its drivers, surge its prices for riders, and crush its competitors if it is going to find a path to profitability. We see particular tension ahead with the rise of autonomous cars, a cost-saving move for Uber that could threaten the livelihood of hundreds of thousands of drivers. The next "buy local"–style movement could very well be "ride human."

For all Uber might do to reform, there are others working on ridesharing platforms that aim to *replace* the worst aspects of its model, creating a harmonious triangle and a more perfect circle, promising a better deal for everyone involved: riders, drivers, and the wider world. Our next story tells of one such effort.

A more perfect circle: The great promise of Ride Austin

Uber broke up with Ligia Friedman by text.

She had fallen in love with being an Uber driver on her second day, earning $300 in a shift—the day of a big football game in Austin. Her job was fun and it helped provide for her teenage daughter as the two found their feet in a new city. So when the city of Austin proposed legislation that would mandate fingerprinting for all drivers—something Uber strongly resisted—she jumped at the chance to help. She attended events and mobilized on behalf of the platform. She was even featured in its promotional materials.

But Uber failed to get its way. So—along with the other big player, Lyft—it abruptly pulled out of the Austin market. It simply texted the decision to its drivers and immediately shut down the app.

It was as if, Mayor Steve Adler told us, Uber had "built the factories and then left, but they turned over the keys. They left a universe of drivers. They left a universe of riders." And indeed, within weeks, a brand-new organization was ready to move in. A meeting at Adler's home had inspired a group of city council people, tech entrepreneurs, and business folks to try a new approach to the same service. And Ride Austin was born.

Ride Austin was created not simply as a *replacement* for Uber, but as an antidote to it. It was an entrepreneurial attempt to create a ridesharing model that would be defined by community, not just by transactions. As TheNextWeb described it: "The app is essentially a clone of Uber, only without the shitty business behind it."

What is behind Ride Austin isn't a business at all, but a nonprofit. It is a ride-share service that serves the community, with a commitment to drive down costs and increase accessibility. As its slogan goes, "Built by Austin, for Austin." It espouses a philosophy that is proudly local, pro-driver, and philanthropic. You can tip the drivers. You can round up to the nearest dollar to support local charities. One hundred percent of the normal fee (and 80 percent of premium rides) goes right into the driver's pocket.

As its former director of community, Joe Deshotel, puts it, "Uber keeps its own power by separating the driver from the rider and putting these digital barriers between them." Ride Austin is breaking these barriers down. It hosts regular driver appreciation events that attract thousands of community members and drivers alike, drawn by BBQ and Texan spirit, and it has raised millions of dollars from local donors who see its emergence—especially in contrast to the extractive business model of Uber and Lyft—as a civic virtue. It even hired Ligia Friedman, whose enthusiasm and expertise made her the perfect head of driver acquisition, acting as champion, therapist, and guru for the driver community.

Ride Austin's efforts to find alignment among its constituents extends to how it interacts with government. Unlike Uber, which hoards its data, Ride Austin shares its ride statistics to help inform transit planning and help design better—and fairer—ways to move people around. In fact, its strategy to scale beyond Austin relies on its promise to other cities of more value and a better partnership. Ride Austin's approach to regulators, as Deshotel tells us, is, "Look, you like the model. You like the platform. We can provide those things with additional transparency and additional cooperation, versus competition."

Ride Austin sees the harmony it can achieve within its triangle and circle as its competitive advantage. It is now committed to making its drivers full employees with benefits—creating reliable jobs. It has plans to work in underserved areas to increase mobility and to encourage more actual ridesharing, by reducing single occupancy rides. It is also thinking through a proposal to use some of the fees from its luxury ride options to offset the costs to low-income riders. Think of it as redistributive ridesharing.

The unique situation in Austin—the Uberpocalypse, as it was known—and Austin's progressive culture offered up a rare chance for an initiative like Ride Austin to emerge. Yet its story may prove quixotic. A new statewide deal has brought Uber and Lyft back to town, leaving Ride Austin's future in serious doubt.

Still, the big lesson of Ride Austin is that we can imagine a radical alternative to Uber and glimpse how much better off we might all be for it. Ride Austin has shown that its service is viable, both in terms of providing rides and in terms of its finances. By spring 2017, it was closing in on profitability, had hit its millionth ride, and had 5,100 drivers delivering 70,000 rides a week. And it achieved all this with great community goodwill and a bank of trust and satisfaction among its key players.

The conventional wisdom is that network effects at work around platforms like Uber will lead inexorably toward a national and perhaps global monopoly or duopoly emerging (and that is certainly Uber's goal). Yet, as Deshotel reminded us, 90 percent of ridesharing is purely local. It is a service fairly well suited to a city scale and a more communitarian bent, especially if the incentives of drivers, riders, community, and government can all fall into alignment.

Going further, it is not absurd to imagine a Ride Houston, or a Ride DC, or even a Ride São Paulo. Using shared technology and branding, local versions of ridesharing apps might even start to assemble some kind of "federated platform," with different communities owning and tending their own patches of the same ecosystem.

THE CIRCLE TEST: BUILDING NEW POWER COMMUNITIES THAT WORK, INSIDE AND OUT

The dynamics of new power communities aren't important just because they help us understand why Reddit might revolt or Uber might combust. They also matter because these platforms are having far-reaching effects on all of us.

Facebook is the world's most populous new power platform with an increasingly dominant hold on global audiences. On a micro scale, it presents a challenge for the local nonprofit or school that wants to reach potential students or donors but must pay Facebook to "boost" its post's reach. And it presents a macro-challenge challenge for democratic discourse itself when Facebook, via its opaque algorithm, gets to choose who sees what and when. It can be tempting to think of social media sites as merely virtual clubs or cafes, and less consequential than the things that unfold in parliaments or on battlefields. But in fact platforms like Facebook are increasingly shaping and eclipsing what happens in these domains.

To take one famous example, we now know the crucial role Facebook played in the 2016 U.S. presidential election, both as the primary means by which "fake news" and misinformation was distributed and spread sideways, and as the vehicle through which Trump campaign affiliates conducted highly sophisticated targeting of millions of people in the days before the election—acting to increase his turnout and depress his opponent's. Mark Zuckerberg's initial reaction to these claims was incredulity: "Personally, I think the idea that fake news on Facebook—of which it's a small amount of content—influenced the election in any way is a pretty crazy idea." Five months later Facebook reversed course and conceded it had played a role. And the news kept dripping out: by September it admitted it had sold ads to Russian-backed operatives trying to manipulate American opinion. By November, we learned that these ads reached more than 150 million Americans.

When we think about the effect of vast new power models like Facebook and Uber on the wider world, we should apply what we call the "circle test." The circle test asks us to consider the impacts of the platform, both on the actors inside its triangle and on those in its wider circle. For Facebook, this test asks some tough questions. What is its effect on the ability of traditional media to carry out their public interest role? What is the psychological impact of Facebook, as the scientific evidence mounts that the more we use it, the worse we feel? What are the ethical implications once it has billions more in its reach (especially relevant as Zuckerberg attempts to bring Facebook-enabled—and possibly narrowly limited—internet access to places like rural India)? Should the "public square," as Facebook is increasingly becoming, be privately owned and controlled?

It isn't just policymakers, academics, and ethicists who should be thinking about applying the circle test. It's us, as participants and super-participants. These platforms are empty vessels without us, and we determine whether they thrive or falter. As we choose whether or not to participate, we have an obligation to consider more than whether new power communities delight or make things easier for us, but also whether they are helping or harming the world all around us.

THE PARTICIPATION PREMIUM

THE EVER-EXPANDING UNIVERSE OF STAR CITIZEN

If the $64 million "stretch goal" was hit, then the space pets would be released.

Potential pledgers were promised that, one day in the future, they would have access not only to "traditional terrestrial options," but exotic animals of the type you might expect if the year was 2946 and you were witnessing the swift decline of intergalactic capitalism.

But we are getting ahead of ourselves.

The space pets came only $64 million into what might end up being a $200 million story—the story of the biggest crowdfunding campaign in history. And there are different predictions as to how it might end—either as crowdfunding's greatest cautionary tale or its crowning achievement.

The right place to begin is before the first dollar was raised, back at the inception, when a new universe exploded onto the scene.

The Big Bang

When legendary game developer Chris Roberts announced he was returning to duty after a ten-year hiatus, gamers rejoiced. The cre-

ator of the classic 3D combat space simulator Wing Commander
was revered, especially by the PC gaming community, which had
felt increasingly unloved as gaming moved to tablets and consoles.

And Roberts had very big plans.

He did not simply want to build a game. He wanted to build Star
Citizen, "a living, breathing science fiction universe with unparal-
leled immersion" allowing for unlimited gameplay. Those not in the
gaming world might miss the scale of ambition here. Roberts was
not only going to resuscitate the PC game, he was going to build
a universe—ever expanding, seemingly limitless, and, most critically, shaped by the actions of its inhabitants. Changes in the 'verse
would be based on players' actions, so if a "Citizen" were to find
a "jump point" into a new system it would be named after him. A
player could design his own spaceships (and maybe even sell them
to other players). He would be master of his destiny.

And with Star Citizen, Chris Roberts was doing more than setting out a vision for what he might build. He was conscripting the
crowd to join in.

The universe emerges

When Star Citizen was first a glint in Roberts's eye, he tried to get
traditional publishers and venture capitalists to fund it, but didn't
get enough interest. This was the age of Angry Birds and Candy
Crush. His epic reimagination of the PC space sim (or, in plain
English, a simulated reality game set in space that you play on your
personal computer) was not turning heads.

So he tried something else.

Roberts took a prime speaking slot at the 2012 Game Developers Conference in Austin and announced his comeback project. In
an hour-long presentation, he unveiled his vision for Star Citizen,
showcasing a range of dazzling prototypes for the game. As he
walked the audience through new worlds, playing a prototype as
he spoke, he came across as the 'verse's own biggest fanboy. Viewers

got into the cockpit with him. They scanned the new horizon. On Reddit, one fan watching the live stream said, "As someone who has played every Wing Commander, Privateer, Starlancer, and Freelancer, this is pretty much my dream space sim. I'm fucking drooling here."

But about fifteen minutes before the end, Roberts's tone changed a little. He seemed to lose a little of his confidence and energy as he began to talk about money.

Given the challenges he faced getting investors excited about the PC game, he told the audience, he had decided to open up an element of crowdfunding to "a limited number of people." He promised they would be involved in the development process, stay informed, and play bits of the game much sooner.

Roberts's "ask" was unpolished and underwhelming. He concluded, "If this is something you're excited by, then please come to RobertsSpaceIndustries.com and participate and support us. So that's my pitch, I'm not very good at the pitch side of things."

Just before he left the podium, he was handed a note. He looked at it with a grin and shared its contents with the audience: "Apparently our site is having millions of hits and is overloaded . . . We'll try and get it back up, I guess, when we can."

Cosmic Inflation

Roberts Space Industries' (RSI) servers went down for days, overwhelmed by the traffic. So it quickly switched the campaign to Kickstarter—the crowdfunding website for creative projects—and set a goal of $500,000.

In the first twenty-four hours, it raised $155,270, almost a third of its goal.

Roberts issued a thank-you note that underscored what was driving the interest: "Remember that the more we raise the more content I can pack in from the outset and louder message we send to the world that Space Sims are very much alive." The pledges were more than preorders for the game (which was originally scheduled to be ready in November 2014), they were down payments on an expanded vision, as well as a vote-with-your-wallet cast in favor of both the PC and space sims. It was a compact between RSI and the pledgers that was part speculation, part support.

It was also part ship. The more you pledged, the better the ship you would have waiting in your hangar when the game was released. Sixty dollars would get you the "Origin 300i spaceship ready to fly." At $125, you'd get the "Anvil Aerospace Hornet ready to fly (High-end ship focused towards dogfighting/combat)." At $1,000 you'd be the proud owner of "your top-of-the-line . . . Constellation space-ship ready to fly."

Interest kept surging, as did the dollars. And the more that was pledged, the more interactive and dynamic the promised universe became. RSI quickly launched a hub to encourage fan-generated content. Called "Spectrum Dispatch," it provided the "lore" of the universe for all to riff off. Fan-generated content is a well-established feature of gaming, but to have it so prominent so early in develop-ment was something new.

In less than a month, pledges had reached $2 million (split between Kickstarter and the now repaired RSI site), the magic number needed to make Star Citizen "a go." Roberts marked the moment with a bullish note to the community recognizing "the sto-ries and the creativity I've encountered interacting with you in the comments and on the message boards and elsewhere."

Rather than taking a bow at this point, having clearly demon-strated the public interest in the game, Roberts kept on fundraising, announcing a series of "stretch goals" that would allow the universe to keep expanding. Each goal hit would unlock new benefits for backers, including more ships and equipment for players, even "life-time ship insurance." (That's right: those who pledged were insuring themselves against the future need to pay for insurance to cover any damage that their conceptual ship might incur if and when the game was finally released.)

And Roberts kept stretching.

By mid-November, Star Citizens had pledged $3.5 million and

were rewarded with the possibility of cockpit decorations like fuzzy dice, bobbleheads, and "nose art."

At $6 million raised, Star Citizen would come with a full orchestral score.

By November 19, after a last-ditch twenty-four-hour live stream to push the campaign to a close at a staggering $6.2 million, a triumphant Roberts announced, "The battle is over and we—PC gamers, Space Sim fans, WingNuts, Lancers and the rest—have won." There was talk at this point of taking down the funding tracker and calling it a day.

But the demand was still there, and who could resist?

In June, RSI celebrated the $10 million mark with a new ship launch, a twenty-four-hour live stream, and a call to arms for the community to share the success and recruit their friends. And it ratcheted up engagement for Citizens, with more video, more access to developers, and multiple chances to share their views.

To mark Star Citizen's first anniversary, RSI celebrated with "CitizenCon," a twenty-four-hour live stream. It generated $2 million in a week and RSI hit $23 million, the number at which Roberts felt the game could be made entirely with crowdfunding.

And he just kept on stretching.

The $50 million mark triggered RSI's commitment to hire linguists to create alien languages to "increase immersiveness." At $64 million came the space pets.

At CitizenCon 2015, this time held in Manchester, England, Sandi Gardiner, senior RSI executive and wife of Chris Roberts, opened the show by reading a personal and emotional letter of tribute to the community. By this point fans had created over 260,000 conversation threads and five million replies just on the *official* site. Gardiner spoke of the deep human bonds this imagined game had forged. "Please raise your hand if you have made at least one heartfelt connection with another human being while being part of this project in any shape or form."

By December 2015, in just a little over three years, RSI had raised a total of over $100 million from over a million Citizens, setting the world record not just for gaming, but for crowdfunding in general.

All for a game that didn't yet exist, and was already more than a year behind schedule.

UNDERSTANDING THE "PARTICIPATION PREMIUM"

What the story of Star Citizen shows us—in both amazing and worrying ways—are the dynamics of how people are spending, raising, and investing money in a new power context.

In the twentieth century, we tended to think of our transactions in one of two distinct ways.

One was the basic economic exchange. You purchase a product—for instance, a refrigerator. You have a very clear expectation that when you order that fridge, a microwave won't arrive instead. If the fridge shows up and it doesn't work properly, you expect a return or a refund. When you're evaluating the price of the fridge you weigh product features and intangibles like brand (you might pay a premium if the manufacturer can convince you its appliance is a "luxury" product). The same logic applies if you're a traditional investor. You're engaged in the transaction in order to earn an expected rate of return, taking risk into account.

The other kind of transaction was more altruistic. As a donor to Oxfam, for example, you don't expect a financial return on your contribution, or really anything other than the intangible benefit of knowing you are helping people (and, in some cases, being recognized for having done it, like having your name listed as a patron). This kind of spending offers, above all, a sense of higher purpose.

In recent years, we've seen these two kinds of returns—the economic and the altruistic—being blended more and more often. Think of impact investing, which promises investors they can make good returns *and* do a little good for the world. Or think of the countless brands that promote themselves as causes—Patagonia, Toms, or Warby Parker, for example. This isn't a new phenomenon—for years, public television has incentivized donations with the promise of free swag—but today it is becoming much more mainstream.

Star Citizen combined "something of value" with "sense of purpose" rather well. Those pledging are expecting that the company will, at some point, actually deliver the game they have purchased. And the value of that product is increased by the sense of higher purpose that the game delivers.

It is no coincidence that the language used around the 'verse could come straight from the NGO campaigning playbook. Star Citizen is constantly cast as the savior of the PC gaming world, in the face of competition from the dark forces of the gaming console Establishment. Those taking part are addressed as PC philanthropists, much as if they were working to free Tibet. Here's a good example of Roberts's proposition to his crowd: "Several years from now, when you are surrounded by your loved ones, and they ask you what did you do during the battle for Space Sims and PC games, you can look them in the eye and say; I helped make Star Citizen."

The sum of these two returns is powerful. But Star Citizen offers something else too, which we can think of as a multiplier: the opportunity to participate. From the start, Roberts's sales pitch promised that Citizens would be able to have a direct impact both on how the universe was going to be built, and on how it would develop once it emerged. The persistent message is one of heightened agency, the sense that Citizens' choices matter and could change the fate of the galaxy.

In the words of one Star Citizen backer, you effectively "pay to dream." By becoming a Citizen, you get your hands on a universe you can shape—one where you can debate the best way to defeat the Vanduul (the archenemies of the game), dissect the designs for the newest ships, or create your own fan fiction stories of how the universe turns out, all with a committed group of like-minded people who don't think you're weird for doing so.

The value of this participation is so great that Citizens are gaining a return on investment even *before* the game exists. As Star Citizen backer powertowerpro put it: "I have already got my money's worth ... I have watched 100's of hours of video content on the making of Star Citizen. I have learned how games are made, 3D ships are built (I am a 3D modeling hobbyist), Q&A, and many other interesting things (some great fan fiction to boot!). That alone is worth my pledge."

When these returns come together, they create what we think of as the "participation premium." Star Citizen offers *something in return* in the form of the promised game and ships and *higher purpose* in joining the mission to revive the PC sim. *Participation* supercharges both: it offers you citizenship in a lively community of fellow dreamers, and even offers the chance to change the game itself. Whatever you're "selling" today, there is great advantage in providing all three returns. To lay this out as an unscientific (but useful) equation, think of this as:

$$(\text{Something in Return} + \text{Higher Purpose}) \times \text{Participation} =$$
$$\textit{Participation Premium}$$

The great power of the participation premium is that it can lead to a decoupling of material value and price. The rewards you get as a backer or participant are not simply tied to the value of the product involved. They are more complicated—and potentially much more significant, for both "seller" and "buyer." This effect is particularly stark in the case of Star Citizen, where there is *literally* no product yet.

In China, the smartphone vendor Xiaomi has used similar dynamics to grow its market—and has been rewarded with a $45 billion valuation.

DIALING UP THE POWER OF THE CROWD

In 2015, powered by its enthusiastic fans, Xiaomi set the Guinness World Record for most smartphones sold in a day—2.1 million units. In just a few short years, the start-up had become the leading smartphone company in China and the third largest in the world.

From its launch in 2010, Xiaomi was committed to offering high-quality, high-spec, low-cost phones. Yet it aimed to be more than a producer of desirable items. Unlike the competitors it was planning to take on, Xiaomi would sell direct to market, not via distributors and third parties. Its relationship with its crowd would be the factor that would make it stand out.

The translation of Xiaomi is "millet," a sneaky reference to the Chinese colloquialism for "revolution." Its followers—the "Mi Fans," as they became known—took up an identity as part of a friendly, countercultural, and quirky revolution (but certainly not a political one), and built up quickly via savvy social media tactics and smart brand artifacts like cute mascots and funky T-shirts. Even its competitors have referred, with respect, to the levels of "idolatry" Xiaomi has inspired in its customers.

Xiaomi feeds this idolatry with the annual Mi Fan Festival—a vast tribute to its fans and a huge sales opportunity that offers a raft of options to engage the company, from offering feedback, to playing new online games, to getting rare deals. Throughout the year, regular "popcorn" events take place all over the country, each bringing together hundreds of its super-fans to connect with the company (and eat a lot of popcorn).

An early 2012 interview with co-founder Bin Lin showed the company had already developed a community of two million people on social media. Crucially, these people were not simply followers; they were active participants in the development of products. Unlike Apple—another brand with a fervent fan base that Xiaomi is often compared to—Xiaomi's instinct was to open up the development process, not shroud it in secrecy. And its fans weren't just voting on which features they liked, either; they were encouraged to create their own designs for the phone's user interface. An article in the *New York Times* told the story of one super-fan—twenty-four-year-old Han Yu—who volunteers for Xiaomi as if it were Habitat for Humanity. He tests the user interface for bugs and works as a moderator on the company's forums. His payoff? He has made friends on the company's platforms and even had one of his product ideas adopted by Xiaomi. "I really enjoy the sense of participation," he says.

By 2015, around the time of its incredible $45 billion valuation, Bin Lin could speak of *40 million members* engaged in Xiaomi's user forums. Wanqiang Li, another co-founder, reported that its smartphone forum receives one to two million interactions *every day*. Each week, on what is called "Orange Friday," Xiaomi offers an updated version of the user interface and asks the community for feedback. As Li describes it, this is like "having a research and development team of 100,000 people." Company lore has developed to celebrate the stories of the creativity of its users: like the fan who had an idea for an improvement to the flashlight function when trying to get home a little drunk one night.

The work of behavioral economists Michael I. Norton, Daniel Mochon, and Dan Ariely helps us understand the dynamics behind companies like Xiaomi. They have identified what they call the "IKEA effect," a tendency of people to place a higher value on self-made products. Their key observation was that when people put together furniture from a self-assembly kit, they tend to seriously overvalue their (often poorly constructed) creations. To test whether the phenomenon holds more broadly, they designed an experiment with a group of people directed to fold origami swans. They asked those who did the folding—"the builders"—and others who were not involved—"the non-builders"—to place a value on the creations: "While the non-builders saw the amateurish creations as nearly worthless crumpled paper, our builders imbued their origami with value." The effect was stark: "Builders valued their origami so highly they were willing to pay nearly as much for their own creations as the additional set of non-builders were willing to pay for the well-crafted origami made by our experts."

Their conclusion? People place a higher value on objects and experiences they are able to shape.

In fact, even when the product is a little absurd and a campaign's purpose is comedic (rather than lofty), a smart invitation to participate can *still* lead to remarkable results. In 2014, Zack "Danger" Brown announced to the world on Kickstarter that he wanted to make a potato salad. His pitch was hardly elaborate: "Basically I'm just making potato salad. I haven't decided what kind yet." His fundraising target was modest: $10. He raised an eye-popping $55,000 from nearly 7,000 backers, with many people tickled by the anarchic fun. Brown's campaign was a roast of the entire crowdfunding genre. Anyone pledging $3 would get a bite of the salad, he promised. In his pièce de résistance, Zack made good on a promise to recognize all his backers while peeling potatoes in what became an epic, nearly four-hour live-streamed YouTube video. It felt a little like one of those endless Andy Warhol films in which you're watching silent footage in slow motion. Some backers remixed the video to highlight their very own 1.5 seconds of fame. A rather mean-spirited YouTube commenter remarked of the name-reading video: "names of idiots who are so sad and alone they have to pay to be a part of something." (To which the internet might respond: Welcome to the twenty-first century.)

THE NEW (AND OLD) SKILLS OF
MAKING AND RAISING MONEY

To generate <u>funding</u> in a new power world—be it via revenues, donations, investment capital, or loans—you need to develop a fresh set of <u>skills</u>, different from the ones needed to make things happen in the twentieth century. Think back for a moment to how life used to be.

Jeremy's father, Frank, spent most of his career as an independent documentary filmmaker in Australia. To get a film made he'd write a "treatment" (it was too expensive to make a teaser video), carefully cultivate a few powerful bosses inside the all-powerful government film funding body, submit himself to the Kafkaesque administrative process, and then wait, fingers crossed, for his fate to be announced.

Henry's mother, Diann, worked as a children's book illustrator in England. Her fundraising tools were her portfolio and her Rolodex.

Her portfolio showcased her technical proficiency and the range of her work and signaled which other publishers had backed her output. Her Rolodex was critical to landing a big "meeting in London," which could lead to an editor raising a thumb (or not), the key to Diann's livelihood. Even a "yes" began a sometimes endless back-and-forth about requested alterations to her work, which were not always served up with much explanation or logic.

These were both worlds in which no more than a few dozen people really mattered, and to succeed you needed to develop insider knowledge of a labyrinthine process. The great skill was to carefully cultivate (and be careful never to irritate) the powers that be and the "tastemakers." Credentials mattered enormously. You needed to "serve your time" working your way up; you needed to go to the right schools. And once you'd broken into this system you had every incentive to try to keep others out.

What is emerging, whether at epic scale on a project like Star Citizen, or on a micro scale with someone forced to raise money for his own knee replacement surgery, is a new set of funding skills. Here's an overview:

Old Power Funding Skills	New Power Funding Skills
Traditional salesmanship	Universally accessible storytelling
Proven credentials and technical expertise	A compelling personal or group narrative
Ability to develop elite patronage relationships	Ability to mobilize a crowd and key social influencers
Capacity to navigate complex bureaucratic systems	Capacity to navigate complex community dynamics
Creating a luxury/exclusivity premium	Creating a participation premium

Of course, old power funding skills and connections still offer a huge leg up—just try being a woman or person of color, or a non-Ivy League grad, trying to raise money in Silicon Valley. But mastering these new skills can offer a way around closed, patronage-based systems. And increasingly those mastering them are succeeding in getting their films made, scaling their businesses, and accruing economic value of all kinds. Or even, as we'll see in the next story, selling beer.

NEW POWER FUNDING SKILLS IN ACTION: THE INTOXICATING TALE OF BREWDOG

BrewDog is a Scottish start-up that has revolutionized its industry—and made it big—through bold and savvy experiments in how to raise money, raise a crowd, and brew beer.

It was started by two twenty-four-year-olds, James Watt and Martin Dickie, with a modest twenty-thousand-pound bank loan in 2007. Their big break came the following year, when the pair entered a blind tasting competition run by the United Kingdom's biggest supermarket chain, Tesco, and their beers came in first, second, third, and fourth. Not bad. The prize was access to millions of UK consumers.

Watt and Dickie are beer-infatuated Scotsmen (aka Scotsmen) with a serious problem with authority (aka Scotsmen), something they have infused into every aspect of the brand. *The Guardian* has recounted some of BrewDog's shenanigans:

"It has, among other antics, driven a tank down Camden High Street; named a beer after the heroin-and-cocaine cocktail that killed River Phoenix and John Belushi; projected naked images of its two founders onto the Houses of Parliament; brewed beer at the bottom of the Atlantic Ocean; dropped stuffed cats from a helicopter onto the City of London; employed a dwarf to petition parliament for the introduction of a two-thirds pint glass; and released, for the royal wedding of 2011, a beer containing so-called natural aphrodisiacs such as 'herbal Viagra,' chocolate and horny goat weed, which it called Royal Virility Performance."

Watt and Dickie combine their anarchic marketing and storytelling skills with an obsession with making the best, most authentic, and strongest beer on the market—they're so serious about this that they have PhD-level microbiologists working at their plant in Scotland. They have earned a hugely loyal, cultish following (even as their detractors decry them as insufferable hipsters). They now have more than forty bars around the world, from Manchester to São Paulo, and BrewDog's product is flying off the shelves—it is the fastest-growing food-and-drinks producer in Britain.

BrewDog, in keeping with its anti-establishment ethos and narrative, has taken a radically different path to support its early growth. It has chosen to rally the (drunken) masses rather than court elite respectability. Instead of relying on further bank loans or venture capital, it raised money—tens of millions of dollars of it—from its own consumers, and in the process, has become an early pioneer of crowd equity, with 24,000 of its consumers becoming "co-owners." (The first 1,329 of these investors ended up with a 2,800 percent return when, much later in its story, BrewDog sold a big stake to a private equity group.)

As James Watt describes it, with characteristic but not unjustified bravado: "The mental shackles that have tied down British SMEs [small and medium-sized companies] to the staid and unimaginative traditional methods of raising funds have been untethered. We have ushered in a brave new world." BrewDog's fourth round of crowd equity financing closed in 2016 and raised a record-breaking 19 million GBP from tens of thousands of ordinary punters, or what BrewDog proudly calls "Equity Punks."

To pull off this feat, Watt and Dickie have had to master the skills of community building. As an Equity Punk, you get an invitation to the world's most entertaining annual meeting (an offer that 6,000 BrewDog investors took up in 2015), which is in effect a massive party with big-name bands and endless beer. You also get discounts on the product itself. On its online community platform for investors, Equity Punks can connect with one another, talk craft beer and life, and have their say in how the company is run. The bonds this creates are as strong as the beer it brews. Again, we see the participation premium at work.

Equity Punks may also take part in the Beatnik Brewing Collective, a beer creation platform "completely controlled by our Equity

Punk investors. You have the opportunity to vote on what we brew and dictate our next instalment of bottled anarchy. You are then invited to join us at the brewery for our annual shareholder brew day to brew the beer you created." Shareholders get to choose everything from the level of "hops and bitterness" to the label and name of the beer.

The interesting twist in the BrewDog story—and a real test of its new power funding skills—will be how it navigates its community dynamics now that it has gone from being an insurgent to being something of a mainstream player. For all their capacity with new power skills, the founders had the old power savoir faire to sell a 23 percent stake for $265 million to a private equity group, landing a $1.2 billion valuation. Although James Watt claimed, "We're not going to let the deal go to our heads, but Martin did buy himself a new jumper [sweater]," the deal raises questions about how easily anarchy can live alongside private equity. BrewDog now needs to learn to *blend* power, getting the best from old *and* new to keep crowd and investors content (more on that in chapter 10).

THE BRAVE NEW ECONOMY: THE BIG IMPLICATIONS OF NEW POWER FUNDING

Implication #1: More "bosses" do not equal better governance

In early 2015, Chris Roberts disappeared from the Star Citizen horizon and holed up for sixty-six days in a London film studio with Luke Skywalker and *X-Files* Agent Dana Scully. Some Citizens started to wonder if he had checked out or lost touch with the community he exhorted. "The Chairman," as Roberts is rather magisterially known, had entered another stratosphere.

The game's fundraising success had allowed Roberts's ambition to grow to such an extent that he would now film the live-action sequences for Squadron 42—the game within a game of the Star Citizen universe—with real actors. Mark Hamill and Gillian

Anderson were cast as leads. For good measure RSI signed up Gary Oldman in the role of Admiral Bishop.

Roberts's absence coincided with the news that Star Marine, a first-person shooter module scheduled for release in June 2015, was indefinitely delayed. The disappearance, and the delay, stoked growing fears about the health of the 'verse.

Enter Derek Smart. Smart was a veteran game developer who, as a pledger to Star Citizen, considered himself to be one of Roberts's "922,034 bosses." With Roberts now AWOL, Smart publicly demanded an audit of RSI, a look at its expense reports, and a precise timeline for delivery. He also questioned how Roberts's wife and brother were involved in the company and compensated. As word of his demands spread, some got behind his campaign and started to ask questions of their own. Yet others took a very different view of Smart's provocations. They began to refer to him, Harry Potter–style, as "He who must not be named."

Smart's campaign got some press and some buzz, which only grew when RSI fed the beast by deciding to refund him, stripping him of his "Citizenship."

There was surprisingly little consensus behind Smart's challenge to Roberts's authority. Some Citizens did share Smart's analysis, and wanted more transparency about where the vast sums of pledged money were going. But many others didn't want to invite more oversight. An unscientific but telling poll on the message boards asked participants, "Would you like there to be a publicly available refund process now, given this latest development?" Seventy-four percent of Citizens voted NO.

Some Citizens worried that increased checks and balances would be a fool's errand. Or a distraction from their cherished mission of building the game they love. Those who wanted to think ill of RSI and Star Citizen, commenters wrote, would never change their mind, no matter what evidence was presented. As one Citizen commented, "I'm not sure you understand how this works. You could have three of these [audits] come up clean and people still wouldn't be satisfied."

It took Roberts's return from filming to calm the crowd and put an end to the talk of transparency and accountability. Once back from Britain, he posted a ninety-minute video titled *10 for the Chairman*, in which he addressed the chief concerns of the community, and updated it on his field trip. He wrote a long letter to the community, too, recognizing RSI's challenges, and offering exciting reports of its progress. Together, Roberts's efforts to engage the community and recognize Citizens' agency managed to quell the rebellion.

If Star Citizen is a masterful example of community building, the Smart rebellion reveals that it is also a story of a million people with few rights and (at the time of writing) no "persistent universe" to move into. And, other than sit and hope, or dispatch an irritated tweet, no single backer can do much about it. More bosses do not, in fact, mean more or better governance.

As Smart learned, it is very tough to mobilize a Citizenry of low-stakes investors to question a commitment for which they have very high hopes. A conspiracy of optimism is hard to unravel. As the gaming blog *Kotaku* noted, "Deep down most people, even if they're not backers or fans of space sims in general, want Star Citizen to succeed. Any failure, or perceived failure, would be an astronomical slight against the careers of hundreds, Chris Roberts, but also for crowdfunding generally. After all, we're talking hundreds of thousands of individual backers."

The irony here is clear. These almost-a-million bosses ended up imposing *less* oversight than the old power construct of a handful of key financial backers with a lot on the line; it seems plausible that a more traditional investor setup might be better at pushing Roberts to actually deliver, rather than just selling the promise of cockpit decorations. Similarly for BrewDog, you can bet that the private equity mandarins at TSG Consumer Partners will apply more sober scrutiny to their $265 million investment than Equity Punks do to their low-level stakes.

The dark side of new power funding is that often, in our enthusiasm to be part of something, we don't spend enough time considering the nature and terms of that participation. We grasp for what we might get, and fail to grasp what we might give up.

For better and worse, regulators in some countries are already moving in to protect backers from unscrupulous crowdfunding campaigns: better if they protect us from fraud, but perhaps worse if they squash the risk-taking at the heart of many of the ventures being funded. (Kickstarter describes the riskiness inherent in its projects as a "feature, not a bug.")

Implication #2: Viral ≠ vital

"In July of 2013, our 4-year-old daughter Eliza was diagnosed with a rare terminal genetic disease called Sanfilippo Syndrome-Type A," the GoFundMe campaign began. "In one terrifying instant, we were told that we would have to watch her fade away before our eyes."

This bleak story came with a silver lining of hope. Researchers at the Nationwide Children's Hospital had developed a breakthrough gene therapy treatment for Sanfilippo Syndrome, and, with just $2 million, could run a clinical trial.

Eliza's parents, Glenn and Cara O'Neill, set $2 million as the target for their campaign, more than anything GoFundMe had previously been used to raise. Initial progress was slow: they found it hard to get beyond their own social networks. In desperation one night, Glenn frantically Googled "how to make a viral video" and found his way to Benjamin Von Wong, a young videographer.

Wong and his colleagues dropped everything and flew to Columbia, South Carolina, to stay with the O'Neills for eight days. Together they created a heart-wrenching short video about Eliza told through the distraught voices of her parents. "Hope is a nice word, but we need action," Cara says, appealing to our sense of urgency. It was emotional storytelling befitting its subject matter— with slow-motion footage of Eliza just being a normal, beautiful kid.

The video was the breakthrough the O'Neills needed—it got the attention of mainstream news outlets, attracted 600,000 views on

YouTube, and massively boosted donations. Toward the end of 2015, when the O'Neills were close to reaching their at-first-improbable $2 million goal, a wealthy Texan businessman came on board, pledging to match all donations from that point onward. In May 2016, Eliza was the first child in the world to be given the experimental treatment the campaign had funded.

Eliza's is a crowdfunding success story if ever there was one. A rare disease that was too small and marginal to attract institutional funding finds resources from the generosity of tens of thousands of small contributors, each now invested in Eliza's story. What's not to celebrate?

However, while individual stories of success lift our hearts, the aggregate picture that emerges from the many cases of new power funding is more complicated. At Davos in 2016 an installation full of innovation buzzwords offered what at first glance seemed to be a new and exciting possibility: "What if public infrastructure was funded by the crowd?" Harvard law professor Jonathan Zittrain took a photo of the installation and added the caption, "Or as some would call it, 'taxes.'" In an old power world, he reminded us, many of the things that are now being disaggregated and crowd-funded were centralized—we paid our taxes and asked governments to make rational decisions balancing society's different needs and priorities. Taxes are the antithesis of crowdfunding: we get almost no agency in deciding how funds are spent (other than a vote every few years to pick the government that makes those decisions); we get no "feedback loops" that give us that visceral, rewarding sense of having been part of building something; we have no personal connection to an individual story or challenge. It's no surprise, given the deeply unrewarding "user experience" of being a taxpayer, that people feel increasingly skeptical of and remote from government.

But let's imagine that a town did crowd-fund its infrastructure spending, rather than rely on the centralizing mechanism of taxpayer-funded government. Chances are that the most interesting public infrastructure projects with the best stories would attract the most buzz and dollars—the awesome solar-powered underground farmers' market would almost certainly play better than the vital but deeply unsexy project to smooth out the gravel on the local highway.

Environmentalists have known this slightly depressing fact for years. When groups like the World Wildlife Fund raise money from the crowd, they rely almost exclusively on what the sector calls "charismatic megafauna." "Save the panda bear" will always beat presenting the strategic plan to conserve the mountainous regions in central China where pandas live. As crowdfunding models proliferate in areas that used to be considered "public goods," like education, a debate is brewing. Is crowdfunding for things like textbooks helping kids—or actually hurting public education by taking the heat off government? GoFundMe has become the biggest crowdfunding site in the United States in large part because it has become a way for people in distress to raise money for their own basic health care, something that is state-provided in many other countries.

Crowdfunding may also act to reinforce privilege. Eliza O'Neill had a father savvy enough to get a viral video made about his daughter and a mother who, as a pediatrician, had the skills and knowhow to navigate the health-care system. For every true "outsider" story in this new, disaggregated world, we often see the already-powerful proving to be the most capable of whipping up the support of the crowd. Though there is no question that crowdfunding has opened access greatly for people to run their own initiatives and put them to market, it currently lacks any mechanism to ensure that its benefits are evenly shared or efficiently allocated.

LIFE IN 2946: THE FUTURE OF FUNDING

For super-fan Nehkara, engagement on Reddit deepened his connection with Star Citizen. As a self-appointed Star Citizen ambassador and recruiter, he had been writing long Reddit posts to help "people understand the game." Through these posts he met, and teamed up with, another Citizen. They decided to found a media company.

Three days later they launched the Imperial News Network with the mission "to bring the 'verse to life" and cover all manner of Star Citizen–related news and discussion. It became a popular hub for other Citizens and occasionally featured RSI team members as guests.

Though the game itself is long delayed, fans like Nehkara are deeply involved in its development. He tells us about RSI's "public test universe" where "every day, backers are helping them test the game, refine it, fix bugs, and find bugs, actually, more often . . . It's a very collaborative effort to build the game, and that's the feeling that you get. This is my project too, you know?"

Nehkara sees his role in Star Citizen as making a small contribution to an epic project: "You're going to build this thing that has never been built before. No one's ever conceived of building something like this before, but you get to put in 3 bricks into that wall. The only reason it stands is because you contributed. That is sort of the feeling that you get."

Nehkara has added a lot more than three bricks. He has spent around $2,075. He owns six ships, has brought at least ten friends into the game, and has bought T-shirts and hoodies. He especially enjoys giving ships as gifts to other players and bringing new people into the community. He calculates his spending on Star Citizen at about $35 a month, less than the $85 a month he spends on his cellphone.

At the time of writing, over four years since the crowdfunding began, and with over $150 million raised, the epic universe that Star Citizen promised is still being built.

Roberts continues to rally his crowd, keeping their dreams alive: "Every day, we get closer to our shared dream of a living, breathing science fiction universe where we can immerse ourselves like never before."

Many will see this as a con, and it well might prove to be. But there is another way of thinking about Star Citizen. Fans like Nehkara have already seen real returns on their investments, paid to them in the currency of community ties and opportunities to contribute and imagine.

Herein may lie the biggest irony of all. The original universe that the fans purchased from RSI—should it ever arrive—will likely never be as immersive, rich, or dynamic as the universe they have already created together.

8

TAKING THE TURN FROM OLD TO NEW POWER

A VERY BRITISH CROWD-JACKING: THE LOST VOYAGE OF BOATY MCBOATFACE

At 2.30 p.m., on Tuesday, May 10, 2016, in the Wilson Room of the Houses of Parliament, a formal inquiry began into what on earth had just happened to one of Britain's ships.

The chair opened proceedings by asking Professor Duncan Wingham, chief executive of the Natural Environment Research Council (NERC), a question about his boss. "Do you think that the Minister is proud . . . or do you think he is going to make you walk the plank?"

To his great credit, and with a level of bombast that was more Donald Trump than British bureaucrat, Wingham declared that the project under inquiry had delivered "an astonishingly great out-come," and that his organization had just become "probably the best known research council in the world."

Though it would end with a parliamentary inquiry, the story of this renegade ship began without much drama or expectation. NERC is a fairly dusty independent agency of the UK government. It is the leading public funder of environmental science, supporting research, universities, and more. It makes grants. It funds research reports. It tackles complex scientific issues.

But in early 2016 it was getting very excited about plans for a new $300 million polar research vessel that was to be "the UK's largest and most advanced research ship yet" and would launch in 2019.

To engage the public in this big moment, NERC created a campaign called #NameOurShip, allowing the public to suggest, and then vote on, the appropriate name for this new vessel. As Professor Wingham said at the time, "We are excited to hear what the public have to suggest and we really are open to ideas." NERC's press release trailed the sober and august names that had already been submitted: Endeavor. Shackleton. Falcon. . . . And then the voting was opened up for a month.

The contest, an earnest and fairly vanilla new power experiment by an old power government agency, soon caught the attention of James Hand, a former BBC presenter. He was not especially moved by names like Endeavor, but was drawn to the more whimsical entries that were popping up. So he typed in "Boaty McBoatFace." He then filled the description field, where he was supposed to justify his entry, with the simple logic that "it's an absolutely brilliant name."

The internet, of course, agreed.

Boaty immediately attracted tens of thousands of votes. The NERC website crashed from all the enthusiasm. Three days after the contest launched, Hand tweeted out a very British apology: "I'm terribly sorry about all of this, @NERCscience."

But this was only the beginning. Soon stories started to appear in the UK news media about the renegade success of the little boat that could, and the large government agency having to deal with the growing wave of social media support lifting up Boaty.

This caught the interest of the global media, with NPR, the *New York Times,* and CNN joining the flotilla. Boaty had become an icon, with the story playing out in the headlines, on TV shows, in pubs, and at dinner tables across the United Kingdom. It had become a cultural moment, at once a celebration of the British sense of humor and a modern parable of Crowds versus Castles. In a fine example of storm chasing, James Hand was approached by the cruise giant Royal Caribbean International and asked to offer his expert ship-naming skills to launch its next vessel. As president and CEO Michael Bayley said, "The people of the United Kingdom know the name of a great ship when they see it."

Boaty was everywhere.

The campaign reached 250 million people online. The #Name OurShip hashtag was used 23 million times on Twitter. The website generated 2.3 million page views.

Of course, Boaty won the vote by a landslide, attracting 124,000 votes, and finishing well ahead of other top ten entries, which included the noteworthy "I Like Big Boats and I Cannot Lie," "RRS Its Bloody Cold Here," and "Boatimus Prime." The whole venture had been what we call "crowd-jacked," separated from its original intent through the whimsy of the crowd.

Yet not everyone found this funny. Science Minister Jo Johnson, it was reported in *The Guardian,* said as the vote closed, "We want a name that lasts longer than a social media news cycle." Rumblings grew that Wingham would exercise his right to overrule the crowd and make the final call.

So it was that on Friday, May 6, Boaty McBoatface was buried at sea.

In a clever—and somewhat cynical—move, NERC declared it would name the ship after the great naturalist, TV presenter, and aging national treasure, Sir David Attenborough. A choice no one could really complain about. To soften the blow further, it claimed that the people's boat would live on in the form of one of the ship's remotely operated subsea vessels, which would be named Boaty McBoatface.

The parliamentary inquiry that followed was partly a slap on the wrist for NERC for all the brouhaha but also a real discussion—in the scientific tradition—of what might be learned from this viral drama.

To provide some insight, NERC brought along an expert witness, Professor James Wilsdon, a social scientist from the University of Sheffield. He gamely defended NERC's Attenborough gambit as a "very elegant compromise," but then managed to tip the room into hysterics when he admitted that he, too, had voted for Boaty McBoatface.

Once order was restored, Wilsdon clarified that he didn't believe that a question like #NameOurShip was the right approach for scientific engagement of the public. He pointed out that if you ask "superficial, low-stakes questions" you'll get a different response than you might on weightier topics.

Under pressure, Wingham conceded that NERC had not had much of a plan for follow-up or future engagement, admitting that "we are only really now beginning to address those longer term questions . . . and sustain that level of interest." Even when thrown a softball question by the chair about whether NERC had thought about creating lesson plans around Boaty for teachers, Wingham—a little snootily—suggested that the primary role of a research council is to work "at a higher level."

One could conclude from Wingham's testimony that NERC didn't have an entirely sincere interest in the crowd's opinion. It liked the idea of the public suggesting something, as long as the something was suitably august. It didn't have any kind of trusted community it could count on to participate earnestly. Nor did it have a long-term (or even a short-term) plan to meaningfully engage the crowd it attracted. And as soon as the naming project started going in a direction the executives didn't like, they defaulted to old power and imposed the outcome they wanted.

At the parliamentary inquiry, those overseeing matters suggested that "the internet" was trivializing science. But, in this case, it is fairer to say that science trivialized the internet. NERC didn't fundamentally need the crowd's help, and it wasn't prepared for the consequences of unleashing its energy and sense of fun. This is a forgivable sin. Many old power organizations are making steps (and missteps) like this as they tentatively move into a new power world.

But NERC's unforgivable sin was in abandoning Boaty at precisely the moment it should have embraced her. The agency had attracted hundreds of thousands of people—and much of the main-stream media—all of whom now had a vested interest in Boaty. Had it leaned in to all that engagement, and proudly smashed the champagne bottle on the hull of *Boaty McBoatface,* it could have built a community that would have delivered for NERC for years. You might imagine a generation of Brits following *Boaty*'s adventures by GPS, schoolkids greeting *Boaty* when she docked in their town. You could imagine T-shirts, mugs, avatars, video games, cartoons. *Boaty* might have become the most participatory vessel in the world, capable of delighting the public but also of providing a portal for more substantial engagement with the scientific inquiry she pursued. A hint of what NERC gave up came in the 2017 spike of press and engagement it got around the first research voyage of the subsea vessel version of *Boaty,* which was ready for the water before the actual ship itself.

NERC had an admirable impulse when it began this misadventure. Like many old power organizations, it sensed a need to engage with a bigger community and open up to the world. But, also like many old power organizations, it didn't go about this in quite the right way.

For old power organizations wishing to move into the new power world beyond the occasional initiative or stunt, there are four fundamental issues to consider as they set out to "take the turn," as shown in the decision tree on the following page. Imagine if NERC had asked itself these questions in deciding when and how to turn to new power.

Strategy: The first question to consider is whether new power really fits with your strategy. What problem are you trying to solve that your old power repertoire can't handle? What opportunity exists in the energy of the crowd that could add real value for your efforts? What new innovations might come from the crowd that would *not* have come from inside your walls? In NERC's case, it wasn't clear that it really needed (or wanted) the crowd to name its boat, given that executives were set on a conventional choice. But part of this equation, too, is the question of what value you can

DECISION TREE
When, and whether, to turn to new power

STRATEGY

Do you need the involvement of the crowd to get a better outcome? Does the crowd need you?

NO   Stick with OP

YES   Consider NP

LEGITIMACY ·······

Do you have enough legitimacy with the people you're trying to engage so that you're not ignored or crowd-jacked?

NO   Stick with OP

YES   Consider NP

CONTROL ·······

Are you willing to cede some control to the crowd within parameters you set, and accept outcomes that are unexpected or suboptimal?

NO   Stick with OP

YES   Consider NP

COMMITMENT ·······

Are you prepared and able to sustain the engagement of the crowd and feed their agency over the longer term?

NO   Stick with OP

YES   Use NP

deliver to your crowd. The crowd cannot be seen as an asset to be strip-mined: any new power strategy needs to consider what's really in it for the people you want to engage with.

Legitimacy: If engaging new power makes strategic sense, the next question is whether you already have—or are prepared to build—trust and credibility in the space you want to engage. This means there are specific communities that you're confident will "show up" for you. This requires more than simply having the CEO join Twitter. When in 2013 JPMorgan, with wounds from the financial crisis still fresh, launched #AskJPM, a Twitter campaign offering career advice to aspiring professionals, its lack of legitimacy was flung back in its face. Its efforts were quickly dispatched by a bemused crowd with questions like "Did you always want to be part of a vast, corrupt criminal enterprise or did you 'break bad'?" One of NERC's challenges was that it didn't have a bought-in, existing crowd to draw on, nor did it do much to engage the people it was meeting for the first time. Its approach was to lower the drawbridge and trumpet "openness," but it didn't have an existing core of supporters who might have tried to redirect the Boaty crowd-jacking.

Control: Meaningfully incorporating new power—and getting something meaningful out of the result—requires a willingness to give up at least some control and accept a range of outcomes, including an answer you might not view as ideal. Otherwise you'll never really unleash people's energy and enthusiasm. This does not mean an organization has to embrace anarchy. But it does mean that once you have carefully structured the ways your communities can engage, you need to be prepared to be surprised if they lead you in an unexpected direction. Had NERC been willing to give up some measure of control, it might have seen the big opportunity that fell into its lap when Boaty took off—but instead it chose to put the genie back in the bottle.

Commitment: Too often, old power organizations see new power as an occasional, peripheral, and intermittent activity. But getting the best results requires a willingness to cultivate the energy and enthusiasms of a community of people over an extended period, which NERC had no plan or intention to do. If new power enthusiasm is simply a flavor of the month, or the passion project of an unsupported mid-level millennial, it is much less likely to succeed. As any seasoned movement builder can tell you, the greatest viral successes and most impactful moments usually come after months and years of consistent investment in your crowd.

If NERC had thought through the above questions, it might have chosen to engage a more specific community of people (perhaps engineers or shipping enthusiasts?) on a more substantive question than boat naming, or it might have made the decision to wait and build a stronger new power base before "going retail," or it might have chosen to go ahead with eyes wide open, ready for something unexpected to happen, and to embrace the storm that ensued.

When organizations confront the big questions posed by the decision tree—and make the decision to "take the turn"—there can be significant upside. The turnaround of the Lego company is a remarkable example of how, with a lot of work and sustained commitment, an old power organization can embrace new power and thrive.

HOW LEGO SAVED ITS COMPANY BY
EMBRACING NEW POWER

A new millennium was beginning, but the Lego company was crumbling.

The seventy-year-old Danish firm had focused too much on its theme parks. It had created a vast number of products, many more than the market would bear. It had lost its connection with its consumers. It was suffering from sales decline and a culture of top-down management. In 2003 it posted its biggest loss ever.

Enter Jørgen Vig Knudstorp, who in 2004 became CEO, the first non-family leader of the family firm. He carried out many of the typical turnaround CEO tasks: cut staff, reduced the number of products by half, sold off peripheral businesses, and refocused the company. Out were the wristwatches. In were the bricks.

Knudstorp represented a big shift in business strategy, but he was closely aligned with the owner and chairman, Kjeld Kirk Kristiansen, who played a key cultural and symbolic role within the company. In an attempt to understand the Lego business better, Knudstorp—often accompanied by Kristiansen—began to attend the fan-organized Lego events that had multiplied across the country. He recalled how important his visit to one such event, Brick Fest 2005, had proved.

"I came with my family, so it was a really nice event. I gave a talk for half an hour with no slides. Afterwards I spent two hours with the fans on Q & A from the stage. It was my breakthrough because I established a personal relationship with them. During the rest of the weekend they came over and talked, took care of my kids . . ."

What he saw in the crowd was great passion. And also great commercial potential. The Lego brand had always been well loved, of course, but the growth of the internet was allowing more dynamic forms of appreciation for the product.

People weren't just getting together at conferences like Brick Fest; fan clubs were beginning to pop up across the world. And nascent online communities were emerging, too. Take MOCpages—now one of the largest online Lego fan communities in the world—which was launched by a single Lego fan, Sean Kenney. When Kenney launched his site, he had been surprised and inspired to see that "visitors from around the world were leaving messages, rating his creations, and sharing their ideas." He witnessed, before the Lego people did, the surging demand among "makers" to share, connect, and build on one another's creations (MOC stands for "My Own Creation"). Today it is a platform for over half a million fan-made designs using Lego sets.

As it tried to dig its way out of crisis, Lego's leadership was sensing a big opportunity in this deeply committed—and increasingly networked—consumer base. But to uncover it, they needed to engage with a group of customers they had historically ignored.

For the first time in their history, the Lego Group decided to get serious about the AFOLs.

Meet the AFOLs

Robin Sather has not gone through a single birthday or Christmas without receiving a gift of Lego. He is in his fifties and remembers getting his first Lego set when he was four. He still has some of those bricks.

As he got older, his love for Lego toys never faded. But this was a source of discomfort rather than pride. Before the internet came along, he thought he "was the only adult that was still into Lego."

He wasn't. Individuals all over world had kept on building into their adulthoods, but they just hadn't found one another. "We all thought we were freaks and all alone and did it in secret," said Robin.

The rise of the internet saw a global network of people like Robin stepping out of the closet (or, more often, the garage). "We all discovered each other, as with many things, a lot of Lego enthusiasts started to form clubs and a community happened." No longer a furtive solo builder, Sather became a proud co-founder of the Vancouver Lego club. He also took on a new badge of honor. He became an AFOL.

An AFOL is an Adult Fan of Lego, someone like Robin who loved Lego sets as a kid and then kept up with their interest in later life. In *Brick by Brick*, David Robertson and Bill Breen reported that this was a group that the Lego company had historically neglected. They were seen as "few in number and not worth listening to." Some saw them in less flattering terms than that. And even those middle managers who saw potential in adult fans were usually rebuffed by leadership.

Yet under Knudstorp and Kristiansen's new direction things started to change, and for good reason. Even though AFOLs were only 5 percent of the total market at this point, they were outspending the average family with kids *twenty to one*. They were responsible for another phenomenon, too. AFOLs would often turn up at a Lego fair, build something amazing, and then 2,500 kids and families would show up. So would local media.

In our new power terms, the AFOLs were the Lego company's super-participants. Invested, passionate organizers, they were capable of generating communities of their own around the Lego brand. Sure, the company's core participants remained the kids, but increasingly Lego's executives would see its economic value, and its community culture, as shaped by AFOLs. The Lego Group's big pivot was to stop treating AFOLs as an underclass of nobodies, or a den of weirdos, and start building structures to respect them, engage them as super-participants, and channel their value for the Lego ecosystem.

Opening up the castle

At the heart of this effort was senior director Tormod Askildsen. He hired the company's first community manager, began to engage seriously with the Lego super-users, and started to identify smart ways to connect the company with its crowd.

This paid off quickly and unexpectedly. One super-user, John Barnes, was an expert in making high-tech sensors. Working with Barnes's firm, the Lego Group made big improvements to their new robotics kit. Another, Adam Reed Tucker, founder of Brickworld Chicago (now one of the largest fan events) and a self-described "guy with an idea," had become passionate about building highly detailed re-creations of great buildings using Lego materials, notably attracting interest and press around his Chicago Sears Tower. Previously, mid-level Lego managers had been serving as AFOL dealers, feeding him bricks to fund his habit, but now senior management engaged. This partnership developed into the hugely popular "Lego Architecture" line, which has created sets for AFOLs to build the world's greatest buildings, from the Sydney Opera House to Falling Water to the Burj Khalifa.

Part of Askildsen's success was his openness to ideas from the outside. Robin Sather remembers putting together a white paper for

Lego that proposed the creation of a formal ambassador network and recognized status for the most talented builders. Lego Ambassador Network now has a full-time staff and an online platform dedicated to supporting and connecting the leaders of hundreds of local user groups from Hawaii to the Philippines to Jakarta. In effect, this is a sophisticated community of Lego super-participants, who work to grow events, develop ideas, and test new markets for the firm. In return, the Lego company provides free kits, funding, marketing support, and, perhaps most of all, a communal identity and legitimacy for its (unpaid) leaders.

The crowd makes the kits, the crowd makes the movies

The "Lego Ideas" platform took engagement with AFOLs even further, bringing to the mainstream an idea that grew from a pilot project from Lego Japan. Put simply, it crowd-sources the creation of the next Lego models. So if you are a fan with an idea for a new set you can submit your plans to be assessed by the Lego community. If 10,000 participants vote to get behind your idea, Lego administrators take it to a formal review. The very best ideas turn into new products. With three seasons each year, Lego Ideas is like *American Idol* for AFOLs. Over a dozen sets have now made their way from the crowd to the shelves. Those whose Lego ideas get to market don't just get glory: they share in the value creation, earning 1 percent of sales revenues.

One winner was Dr. Ellen Kooijman (online alias Alatariel), a geochemist who had never before shared any of her Lego creations with anyone other than her husband. She was frustrated that female Lego figures were in the minority and fell into stereotypical roles (the Lego Group suffers from a heavily male bias in both its fan base and its corporate culture). As a scientist herself she saw the need for "professional female minifigures that also show that girls can become anything they want, including a paleontologist or an astronomer." Women, scientists, and AFOLs everywhere agreed.

After years of cultivation, the Lego community is now a critical piece of business infrastructure, as important to the company as its factories or intellectual property—something that it can count on in large and small ways. The base was critical to the success of *The Lego Movie* in 2014, which was a huge hit, bringing in almost $500 million in worldwide box office returns. When Twitter user @Shiz_ Nit asked, "Is there anyone out there who is super excited about The LEGO Movie and is older than 32 so I can feel slightly better about myself?" Lego staff knew exactly how to respond, replying, "Age Is Just a Number," and encouraging this AFOL to join community engagement efforts around the film. In fact, AFOLs were active in both the development and promotion of *The Lego Movie*. The film's creators invited the Lego community to film and submit their own scenes—and several fan creations actually landed in the final cut.

Looking back on a decade of change, Robin Sather, now one of just over a dozen Lego Certified Professionals, the company's most prestigious title for expert builders, puts a nice wrapper on things. "The Lego company was like Willy Wonka's chocolate factory. It was this massive edifice where delicious, wonderful things came out the front door, but you never really knew how. You had no access inside. You didn't really know where they came from. You didn't know how it all worked. As the internet flourished and a bunch of different things happened, the doors gradually opened up, and we began to see inside. The whole renaissance of the Lego company opening itself up to the community and to the world has been amazing, and it just continues to evolve."

The Lego company was able to say yes to each of the questions in the decision tree, giving us a good sense of why its turn to new power was so successful.

The Lego Group built a **strategy** in which cultivating its new power community was key to reviving its core business—this wasn't just window dressing or pandering. The company's connection with the AFOLs and its wider communities acted as an incredible—and incredibly inexpensive, compared to traditional broadcast advertising efforts—marketing machine (just glance at how many fan-created Lego YouTube videos there are). The crowd also provides it a huge expert innovation resource at very little cost. As Knudstorp has noted, new products deliver 60 percent of total Lego sales each year. Executives also had a clear sense of what *they* could do for the crowd. Yun Mi Antorini, the Lego community engagement director of innovation and development, and an academic who has studied the Lego Group over the years, sums up the company's philosophy toward the crowd: "You're crashing someone else's party . . . What can you bring to the party?" By putting community management at the center of their business and investing in carefully structured incentives to participate, Lego leadership made being an AFOL more rewarding and meaningful.

Lego staff also worked hard to develop **legitimacy** with the crowd. The early efforts of both its owner and its CEO to engage with their previously neglected communities created the right signal. They earned legitimacy with the AFOLs by legitimizing *them,* taking them out of their AFOL closets and putting them at the center of the company's culture and innovation engine. This offers a key lesson: if you want to be taken seriously by your crowd, taking them seriously is a good place to start.

When it came to thinking about **control,** Lego had something of an advantage. The core design of their product relied on their consumers' shaping the product. As Antorini put it, "The Lego Group could never, can never, and should never, as to this day, really influence what users are doing with the product." But leadership should still be credited with an essential insight: that by relinquishing control even further, they would uncover value. By providing smart routes for participation, Lego staff was able to guide its crowd in certain directions, but without insisting on outcomes. The surge of support behind Alatariel's female scientist Lego figures is a good example of how the crowd course-corrected the company, which had neglected to meet this need.

All this progress required a long-term **commitment** from Lego. It went through an evolution over a decade, not a revolution over a season, with a steady stream of initiatives and experiments that shifted the structure and culture of its business and its communities. As Antorini told us, this has *not* been easy. Talking about "win-win" models is easy to do, but striking the right balance between crowd and company is "hard work to make it work on both sides." This commitment came from the very top of the institution—and has never been a stunt.

This thoughtful and deliberate approach delivered ten consecutive years of organic growth and a renewed dominance in its sector. By 2015, Lego had surpassed Mattel as the world's largest toy company and was anointed "World's Most Powerful Brand" by a leading brand strategy consultancy, not least because of the deep passion of its community.

BUILDING A NEW POWER TEAM

At the Lego Group, taking the turn to new power required more than the leadership of CEO Knudstorp. Its family owner, Kjeld Kirk Kristiansen; its chief change agent, Tormod Askildsen; and its community managers and key AFOLs like Robin Sather and Dr. Ellen Kooijman all played distinct roles and worked together to pull off its transformation. And that is true more broadly. As organizations take the turn, four important characters tend to emerge. Here are the four—alongside four false prophets who are often confused with them.

New Power Team	False Prophets
The Shapeshifter	The Disrupter
The Bridge	The Digital Beard
The Solution Seeker	The Problem Solver
The Super-participant	The Stakeholder

The shapeshifter (vs. the disrupter)

The shapeshifter is a new power change agent in old power garb, a figure with unimpeachable institutional credibility who smooths the path to change and sets an example for the timid or resistant to follow. The shapeshifter figure will likely not be the person who executes day-to-day the big structural changes that make transitions happen, but will be the spiritual and symbolic figure who, steeped in tradition, is ideally placed to guide the institution toward a new identity and a new relationship with its communities.

At the Lego Group, the role of shapeshifter was deftly played by family owner Kjeld Kirk Kristiansen, whose enthusiasm for engaging with the crowd was a critical signal to the organization that this new innovation was *not* a dereliction of Lego history or tradition, but a powerful validation of its new power community. His close alliance with CEO Knudstorp offered the right symbolic support for what became a huge shift in culture.

Another example of the shapeshifter, as we will see in the next chapter, is Pope Francis who has used his infallible old power credibility to unleash unexpected change. Contrast the shapeshifter with a much more familiar modern figure in organizations, the disrupter.

The rallying cry of this generation has in many ways been "disrupt!," with traditional institutions being quickly classified as dinosaurs and earnestly instructed in the ways of the future by those with slick spectacles and thin résumés. When former Yahoo executive Guy Vidra was brought in to run *The New Republic,* a venerable old power public-interest magazine that wanted to transition into the digital era, he delivered a notoriously off-key speech to the magazine's writers and editors, as recounted by *The New Yorker:* "They say that there's two types of C.E.O.s," he said. "There's the peacetime C.E.O. and the wartime C.E.O. Not to be overly dramatic about it, but this is sort of a war. This is a wartime period. That just means that we need to change a lot of things. We need to just break shit . . . And it's scary. It's definitely a scary thing to do. But it's also fun: you know, lean up against the wall and break it." Not long after, the editor in chief and most of the magazine's staff resigned in protest.

Leading an old power organization through transition isn't about "breaking shit." It requires a tricky blend of tradition and innovation, past and future. Those efforts need shapeshifters who can show—by example—how to get the best of both worlds.

The bridge (vs. the digital beard)

More than one old power organization has found itself with a "chief innovation officer" or "director of strategic initiatives" parachuted in by a CEO desperate to uncover some magical revenue line, or to serve as public evidence that her leadership *really is* engaging with the new world. But despite good intentions, these people often end up as "digital beards," providing cover for a risk-averse leader and an unchanging strategy, and relegated to the margins of power and influence within the organization. They are often a small department paid to think about the future, resented by the rest of the organization for not doing what is considered "real work." They cut ribbons at openings for 3D printing labs but in reality are often siloed and underfunded.

Instead of the "beard," what organizations really need is a "bridge," that person who can meaningfully connect his organization to the new power world, making the practical "jumps" between old and new power. A bridge's work is structural, above all. At the Lego Group, that bridge figure was senior director Tormod Askildsen. It was Askildsen who had conducted the earliest new power experiments there and was tapped by the CEO when the company was ready to make a bigger shift. He was effective at navigating internal systems and culture, painstakingly built up community management at Lego from one lonely hire to a whole function in the organization, and worked to integrate his team (and the wider community they represented) with the product designers, marketers, and other "incumbents."

To show a bridge at work, meet Helen Bevan, a change-maker

living inside one of the world's most quintessential old power systems.

The National Health Service, the UK's publicly funded health-care system known by all as the NHS, has more than 1.5 million employees. Only four organizations in the world are larger: McDonalds, Walmart, the People's Liberation Army of China, and the U.S. Defense Department.

For more than twenty-five years, Helen Bevan has worked as an "internal change agent," looking to tame this much-loved but often maligned bureaucratic beast. Her role has been to find smart ways to make things better, from reducing waiting times for patients to improving cancer care.

She describes the NHS as "two old power systems inside each other." First you have an organizing model that is very command-driven and top-down. Then you have the clinical hierarchy, with doctors at the top and patients at the base.

As her career progressed in the 1990s and 2000s, the NHS was shaped by its efforts to meet big and demanding national targets. Targets for waiting times. Targets for what happened to people who had a stroke. The targets were crude tools, but they had real value. They ushered in a more accountable culture and delivered some tangible improvements. Yet they often came at a cost. Achieving them frequently had a negative impact on staff morale and patient care. For many the culture of targets was a culture of fear, governed by anxious higher-ups. More and more, Bevan felt, the NHS was "hitting the target, but missing the point."

One of her signature achievements is the "School for Health and Care Radicals" (which has been recently renamed the School for Change Agents). This pop-up school rallies and connects the community she sees as key to transforming the NHS: heretics and innovators who can bring new ideas to the system. The school operates primarily as a five-week MOOC (massive open online course), but has broadened to include online discussions, mentoring, and in-person activities for those taking part. It is free and open not just to NHS employees, but to anyone in the world. Thousands of people participate in the school, whose benefits range from offering peer support to the colleague on the verge of leaving to giving a big boost to the wannabe change-maker who now feels "this has given me the permission to actually say, 'Yes, go ahead and do it.'"

The school brings together much of the thinking that Bevan has championed in recent years. In 2013, she and a young clinician dreamed up NHS Change Day, a grassroots effort to mobilize people throughout the NHS system to pledge one thing they might do to improve life for patients. It offered a lot of freedom and few conditions or rules, encouraging people to choose just one commitment they might make on the same day as their colleagues. It came at a time that the NHS was feeling especially strained, with increasing demands and diminishing funding.

Change Day was a classic ACE campaign, with a clear call to action that connected staff around an inspiring goal and an extensible frame that participants could make their own. Nurses in the city of York pledged to create a mock ward where they would play the role of patients, in an effort to increase empathy. In Leicester, they cleared out old paperwork, dispensing with two hundred boxes of unnecessary literature. In East Cheshire, staff set up a weekly ritual where they would come together to ask, "What have I done to help a patient this week?," guarding against the danger of getting stuck in administrative work rather than engaging with patients. Many people pledged to smile at colleagues and patients more. Doctors pledged to taste the pediatric medications they dispense. Teams hit the streets to engage with the public to ask their opinion on how the NHS might improve. In all, in its first year Change Day generated 189,000 pledges. By year two, there were over 800,000. It was the first major grassroots mobilization inside the system, unleashing the vast agency of frontline staff.

As Robert Ferris-Rogers, director of engagement for a large NHS region, put it:

> Change Day fundamentally changed the way I thought about
> engagement and connection. It was phenomenal to be a part
> of it and the power that was created by commitment rather
> than compliance. The even more inspiring element of this

was that it was undertaken in [one of] the most hierarchical organizations in the world. The breaking down of barriers and liberating staff to make a difference was brilliant—and I know it made a difference to people's lives (both staff & patients).

Bevan's team also catalyzed a fresh approach to treating people with dementia, an arena where "very powerful anti-psychotic drugs [were being prescribed] to people living with dementia to subdue them and keep them quiet." These drugs had harsh potential consequences, leading to strokes and even death.

As Bevan notes, the typical old power approach to a challenge like this would be "somebody at the center at the top of the system would say 'We are going to create a new standard and we are going to instruct doctors not to give these drugs to patients and we are going to instruct pharmacists to make sure the doctors don't give the drugs to the patients.'" She had seen many times the failings of relying on a top-down system alone.

Working with a national charity called the Dementia Action Alliance, Bevan's team used movement-building tactics to start a national effort around the issue, bringing in doctors, pharmacists, nursing home staff, caregivers, and a constellation of other actors. It was "classic new power," in Bevan's words, with "lots of different communities and groups all doing their own thing." Together, in less than three years, they delivered a 51 percent reduction in prescribing antipsychotics to people with dementia.

With efforts like the School for Health and Care Radicals, Change Day, and the dementia initiative, Bevan is trying to channel the energy that clearly exists inside her system, and turn it from occasional stunt to a cultural norm. Her aim is to create a "community of change agents," who operate in what she calls the "zigzag-y" place between old and new power. These people are not necessarily linked by their positions, nor by their specialties, but by a genuine interest in mobilizing their communities to effect better health outcomes.

Bevan is the perfect example of a bridge to new power. Incumbents, whether individuals or longtime institutions, are often ready to roll their eyes at the efforts of bridges like Helen Bevan. The price of occasional public failure can seem higher than quietly managing decline. Yet they are essential figures in organizations wishing to transform.

The solution seeker (vs. the problem solver)

Back in chapter 2, we told the story of NASA's experiment with open innovation: all the opportunities it unleashed and all the troubles it caused. You might remember that the scientists and researchers divided into two groups: one that resisted engaging with the crowd, one that embraced it.

As NASA's experiment with open innovation unfolded, its scientists developed an internal vocabulary to describe the difference between fiercely divided camps: "problem solvers" and "solution seekers." Their archetypes will ring a bell for many when they think about their own workplaces and teams.

The "problem solvers" were those who resisted change. Their identity was invested in their own expertise. They were scientists in the tradition of Galileo, those whose individual genius would light the way.

The "solution seekers" got their name from a heated debate that Hila Lifshitz-Assaf observed, where a highly respected scientist rebuked reluctant colleagues: "Your main responsibility is to seek for solutions and they may come from the lab, from open innovation, or from collaboration, you should not care! You are the solution seeker!"

It was this group of scientists that got creative and serious about the wider community. They shifted the boundaries of their worlds to invite people in. For this group, success wasn't that you *personally* had the answer, but that you were open to experiment, ready to find answers in unexpected places and from unexpected people.

Any team "taking the turn" needs to build a squad of solution seekers. These are the people, typically drawn from the main body of staff, who will become the experimenters and allies of new power initiatives. Investing in this group—and recruiting for it—is key, not just in creating new value for an enterprise, but as a political force.

In old power organizations, all of our resources, training, recognition, and rewards are geared to problem solving. Shifting them is hard work. At NASA, it was the smart work of Jeffrey Davis and his team in their bridge role that encouraged people inclined to be solution seekers. They provided a path for those who were prepared to become a new kind of expert, secure and self-confident enough in their abilities to look outward. They did this with an "opt-in" approach, creating frequent lunch events, sessions with outside speakers, and opportunities for the curious to find out more. Then, as soon as interest started to build and results started to be generated, the team of administrators got themselves out of the picture. They made sure it was the solution seekers, rather than the staff overseeing the initiative, who became the stars of the show. These stars recruited others, and the cultural shift really began to take root.

The super-participant (vs. the stakeholder)

No new power team would be complete, of course, without those people who create huge value in the community: the super-participants. These are the most engaged and active people in your wider crowd—AFOLs like Robin Sather, who set up the Vancouver Lego club, and Bruce Cragin, the semi-retired telecommunications engineer who out-thought a group of the best-resourced rocket scientists in the world. Though his idea was the most celebrated, he was just one of many super-participants who contributed valuable thinking to NASA's efforts.

One of the dangers of transitioning to new power is in seeing the crowd as a distant, amorphous asset—a blurry mass of occasional opportunity. But crowds cannot be approached in the same way that organizations often treat such "stakeholders" as civil society or investors: as external actors who must be managed (and sometimes tolerated) alongside the pursuit of the activities that "really" matter. In contrast, super-participants always *participate*, and they create value by doing so.

A company like Lego has developed such a valuable community of super-participants because it has been prepared to genuinely embrace the world outside its four walls and beyond its payroll. Unlike NERC, which dipped its toe into new power waters and retreated at the first sign of a wave, the organizations who get this right are those who dive deep.

LEADERSHIP

"Thank you for everything. My last ask is the same as my first. I'm asking you to believe—not in my ability to create change, but in yours."

—Barack Obama, January 2017

"I alone can fix it."

—Donald Trump, Republican National
Convention, July 2016

How do we make sense of leadership when the world's leading democracy can elect Barack Obama—and then replace him with Donald Trump? How do we reconcile a world of ever-proliferating "leaderless" movements like Occupy Wall Street and the Arab Spring, but also ever-proliferating strongmen like Russia's Putin, Turkey's Erdoğan, and Egypt's el-Sisi?

This chapter is about leadership in a world of colliding—and overlapping—old and new power forces. It's about the ways new power tools are unexpectedly being used to strengthen and reinforce leaders with deeply old power values. And it is about emerging forms of leadership that aspire to bring together new power tools and new power values in order to meaningfully redistribute power.

To start to make sense of it all, let's take the new power compass from chapter 2 and apply it to how people lead. If we map out both the values and leadership models people are using, we can draw four different leadership archetypes. You will recognize some of these in the news and in your own world.

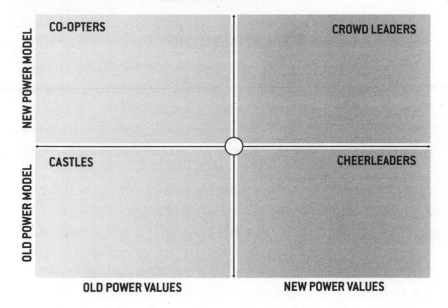

The **Crowd Leader (top right)** combines a new power leadership model with a commitment to, and articulation of, new power values. The Crowd Leader wants to do more than channel the power of her crowd; she wants to make her crowd more powerful.

The **Cheerleader (bottom right)** champions new power values like collaboration, transparency, and participation, but leads in an old power way. He either isn't able or doesn't want to genuinely distribute power.

The **Castle (bottom left)** pairs old power values with an old power leadership model—this is the traditional hierarchical and authority-based model of leadership most of us grew up with, and which is widespread in sectors like the military, business, and education.

The **Co-opter (top left)** deploys a crowd and skillfully uses new power tools and tactics—but does so in the service of old power values, and to concentrate power for himself.

BARACK OBAMA AND DONALD TRUMP:
A STUDY IN CONTRASTS

The charismatic Crowd Leader: The campaign of Barack Obama

When Barack Obama ran for president in 2008, he was the ultimate Crowd Leader. His iconic chant was "Yes We Can!" He called for his supporters to look to one another, not just to him: "*We* are the ones we've been waiting for. *We* are the change that we seek." While Hillary Clinton told stories of her own agency, that she was "ready on day one," Obama was talking about building a movement and scaling up the community-organizing skills he once used to help the disenfranchised on the South Side of Chicago.

His was a highly participatory campaign—in fact, it was a stunningly ambitious and successful use of new power tools and tactics. The digital hub of it all was MyBarackObama.com ("MyBo"), a sophisticated (for its time) platform to help people organize, volunteer, and raise money. These online efforts helped mobilize record numbers of volunteers under the motto "Respect. Empower. Include." The campaign got to such scale because it gave volunteer leaders real responsibility and incentives to recruit and develop their own local teams. This explosion of new power energy helped the relatively unknown Obama eclipse the old power donor apparatus of the Clinton campaign in the closely fought 2008 Democratic primaries, raise hundreds of millions in small donations, and win resoundingly in the general election.

For all Obama's talk of "we," however, his campaign still relied heavily on the magnetism, charisma, and symbolism of Obama himself. Shepard Fairey's iconic "Hope" poster wasn't of a crowd of people, it was of Barack Obama. He *was* the hope. Yet he managed to then use that energy to create a genuine grassroots movement and to champion values that made *others* feel powerful.

Obama ran a campaign that was at once highly participatory and highly structured. Everyone had a lane in which to participate; as an Obama volunteer, you had some room for creativity and agency but you also had a clear brief and accountability, which were summed up in the epic 280-page organizing manual the campaign developed. As the legendary field organizer Zack Exley described it after visiting Obama's field offices in Ohio in 2008: "Other recent attempts have failed because they were either so 'top-down' and/or poorly-managed that they choked volunteer leadership and enthusiasm; or because they were so dogmatically fixated on pure peer-to-peer or 'bottom-up' organizing that they rejected basic management, accountability and planning. The architects and builders of the Obama field campaign, on the other hand, have undogmatically mixed timeless traditions and discipline of good organizing with new technologies of decentralization and self-organization." Think of this as an exercise in "blending" power, an art we will describe in chapter 10.

All that structured participation worked superbly in an election setting, when goals and an end point were clear, and feedback loops—doors knocked on, voters called, funds raised—were closed.

Across the nation, people like volunteer Jennifer Robinson—whom Zack Exley described in his account of the Obama campaign—found the experience transformative: "I'm a different person than I was six weeks ago . . . I've realized that these things I've been doing as a volunteer organizer, well, I'm really good at them, I have a passion for this. I want to continue to find ways to actively make this place, my community, a better place. There's so much more than a regular job in this and once you've had this, it's hard to go back to a regular job."

But, as we shall see, after the election, this highly structured and directed approach to involving people became more straitjacket than supercharger.

From rent-a-crowd to intensity machine:
The campaign of Donald Trump

"Wow. Whoa. That is some group of people. Thousands . . . This is beyond anybody's expectations. There's been no crowd like this."

Donald Trump began his presidential campaign by buying himself a crowd. He reportedly hired a casting agency and paid actors $50 each to show up with "homemade" signs and cheer him on as he descended the elevator at Trump Tower. He started that speech by noting with surprise just how many people were in attendance, and then offered a rambling *tour d'horizon* of the epic failings of America and his unique capacity to put things right.

China was beating America. So was Japan. So was Mexico. The situation in the Middle East was a disaster. Obamacare was a disaster. Obama was a disaster. "I beat China all the time." "Nobody builds walls better than me." "I've done an amazing job." "I don't have to brag. I don't have to." "We need—we need somebody—we need somebody that literally will take this country and make it great again."

Many rolled their eyes, dismissing him as they would their mad uncle muttering obscenities at the television. Others took notice, the brazen self-confidence, a clear list of enemies, and the money-back guarantee of a better life all hitting their mark.

As we know now, this rent-a-crowd would grow into a real one, capable of destroying the established political wisdom, embarrassing the pollsters and pundits who had promised Trump had no hope, and vaulting him into office.

While both Trump's campaign and Obama's in 2008 showed a mastery of how to work up a crowd, there was one crucial difference: Obama's approach to participation was highly structured and crafted, where Trump's was unstructured to the point of anarchy. There was no reliance on a carefully organized field army. It was the difference between a weighty organizing manual and a 140-character tweet. Whether this was strategic genius on Trump's part or just disorganization (or a little of both), this gave Trump's supporters the freedom they needed to make his message their own.

Trump caught fire in no small part because of his intuitive grasp of how to build a movement in a new power world. As we touched on in the introduction, Twitter turned him into the leader of a vast, decentralized social media army that took its cues from him—and in turn fed Trump new narratives, conspiracy theories, and lines of attack. It would become a deeply symbiotic relationship. On the night of Trump's election victory, the bulletin boards of 4chan, a Reddit-like social network that attracts mainly young men and which prides itself on extreme views and provocations, lit up. The white supremacists who had championed Trump's candidacy couldn't believe their luck. As Abby Ohlheiser recounted in the *Washington Post*: "'I'm f——trembling out of excitement brahs,' one 4channer wrote Tuesday night, adding a very excited Pepe the Frog drawing. 'We actually elected a meme as president.'" Pepe the Frog is the green, human-like frog developed by Matt Furie for his comic book series more than a decade ago. The 4channers appropriated the once-innocent Pepe into the symbol of the alt-right movement, and soon added a mop of orange hair to his image so that Pepe would embody Trump himself.

During the campaign, rather than distance himself from these meme-making extremists, as the typical politician would do, Trump empowered them. He signaled repeatedly to his most extreme supporters to keep at it, and indeed to go harder. He retweeted white supremacists who spread even more overt versions of his own (barely) more camouflaged race messaging. In one instance, he retweeted from a now-deleted account (@cheesedbrit, whose avatar was a neo-Nazi symbol), the false and racially provocative claim that 81 percent of white murder victims are killed by blacks; the real number, according to FBI statistics, is around 15 percent. When challenged, Trump refused to apologize and simply said he couldn't check all the facts he retweeted. The message was clear: I'm not going try to rein in my supporters.

He went further. His unambiguous message to his crowd was

that he'd have their back no matter what they did. He notoriously offered to pay the legal fees of a white supporter who punched a black protester at a campaign event during the primaries. He told his crowd, "If you see somebody getting ready to throw a tomato, knock the crap out of them, would you? Seriously, OK? Just knock the hell—I promise you, I will pay for the legal fees. I promise. I promise."

The secret to Trump's success was not simply his own unrivaled ability to grab the attention of both social and mainstream media. His message also moved—and morphed—without him via a vast network of people who took it and made it their own (a classic ACE effort). These were groups that had been activated in recent years by causes like the Tea Party, the gun rights movement, and the nascent alt-right. Trump took that existing energy, dialed it up to fever pitch, and transformed his campaign into what we call an *intensity machine.*

Throughout the campaign, Trump had significantly higher levels of social media engagement and sideways content creation from his supporters than his opponent Hillary Clinton. In national opinion polls, his supporters were about ten points more enthusiastic about him than Clinton's supporters were about her, even as his favorability ratings overall trailed Clinton's by about ten points. Positive sentiment about Trump on social media also outstripped Clinton's by ten points in the critical final five weeks of the election, according to 4C Insights.

It tells us something interesting about the role of new power in the 2016 election that social media analytics firms like 4C, Spredfast, and Socialbakers proved better at predicting the strength of Trump's support than traditional pollsters. 4C was the same firm that correctly predicted the Brexit vote: conventional wisdom among elites and even betting markets predicted a remain vote with 90 percent odds on the day of that vote, similar to the odds they assigned to a Clinton victory just hours before her defeat. Strikingly, according to these social media analysts, the moment during the campaign that expanded Trump's social media following the most was his campaign's supposed nadir: the release of the *Access Hollywood* tape in which he boasted about pussy-grabbing. This moment, with much of the country lined up against him, caused his supporters to rally around him like at no other time in the race.

Donald Trump: The Platform Strongman

Following Trump's election, George Orwell's *1984* saw a surge of sales. The tale of a totalitarian dystopia, overshadowed by the all-powerful figure of Big Brother, rode a "Trump bump" back to the top of the best-seller list.

Yet the comparison between Big Brother and Trump is not quite right. While Big Brother wielded his power through mass conformity and a fierce resistance to individualism, Trump became stronger by supercharging individual agency, taking the reins off and championing unorthodox, previously socially unacceptable behaviors. His drug of choice is not one that subdues the populace, but one that excites it. And he thinks about media differently, too. While the "telescreens" in 1984 serve as ways to standardize and contain citizens' actions, Trump's platforms—especially Twitter—serve to individualize and liberate. He seeks control of the conventional media and the establishment not by co-opting it, but by undermining it, seeing it overrun by a battery of constant attacks from his crowd. He has no need for a Ministry of Truth to ensure that one state-sponsored opinion is held by all. In fact, he revels in the instability of countless truths. The more the merrier.

Trump is an example of what we call the Platform Strongman: a leader who co-opts a digital crowd and deploys new power in order to advance largely authoritarian values. He rallied a highly empowered, anarchic, digitally savvy movement of people while he paradoxically championed a more orderly and severe, rather than more free and open, America. His campaign speeches painted a jarring picture of a chaotic and violent country, magnifying and vivifying threats from outside and in, and presented himself as the only man strong enough to take that threat away. Trump also promised to restore the "natural" order. By implication, that new order would re-elevate Trump's most intense supporters—less educated white men—at a time when many felt they were losing status. And it's in the context of these authoritarian values that all of Trump's chest-thumping, self-aggrandizing, and aggressive behavior might have appealed so strongly to some Americans, and so viscerally repulsed others. His promise to restore law and order (a coded attack on people of color at a time of historically low crime rates) and his focus on "enforcing the rules" of the immigration system all had a powerful draw.

As early as January 2016, researchers had identified "the one weird trait that predicts whether you're a Trump supporter"—and it wasn't your gender, age, or religion. It was whether you had an authoritarian value set. Pollster and researcher Matthew MacWilliams found that Trump was the only candidate in either the Democratic field or the sixteen-person Republican field whose support among authoritarians was statistically significant. MacWilliams hypothesized, nearly ten months before Trump's unexpected victory, that "in a general election, Trump's strongman rhetoric will surely appeal to some of the 39 percent of independents in my poll who identify as authoritarians and the 17 percent of self-identified Democrats who are strong authoritarians."

The Platform Strongman is a particular species of the Co-opter archetype in our leadership compass, and it is where we see such a big difference between Trump's campaign and Obama's. This combination of an old power, authoritarian value set and a sideways, unstructured new power model lies behind some of the most potent and dangerous leadership models in the world today. This is precisely the combination that ISIS deploys—it offers a return to order and certainty (namely, a brutal medieval order) but spreads that message with savvy, decentralized control.

Slipping from Crowd Leader to Cheerleader: Obama in office

"If you want to know how I'll govern," candidate Obama promised, "just look at our campaign."

President Obama took office with the promise that the more than 14 million Americans, like Jennifer Robinson, who had contributed in some way to his victory might now enter government with him.

Exley had laid down the marker before the election: "Obama must continue to feed and lead the organization they have built—either as president or in opposition." But that was not quite how it panned out, as we've noted. Obama campaigned as a Crowd Leader, but he governed as a Cheerleader. While the rhetorical torch for new power stayed lit, he failed to build a genuine movement to help him govern, to help elect his successor, or to create a sustained and local grassroots to help his political party win up and down the ticket. Obama left office with some major legislative accomplishments and relatively high approval ratings. He had governed for eight years with no major scandal, and his original supporters still had a strong emotional connection to him. But his political opposition had won back the presidency, controlled both houses of Congress, and dominated politics at the state level.

The missed opportunity for Obama was that he had no real transition plan for his crowd. All the energy and commitment that people felt had nowhere to go next. As *Rolling Stone*'s Tim Dickinson has reported, in the critical two months after Obama's election victory, Obama's much-vaunted campaign largely went dark, a decision the political technology observer Micah Sifry called "criminal political negligence." The administration then made the fateful decision to fold its organizing infrastructure—renamed Organizing for America (OFA)—and its more than 13 million members into the Democratic National Committee, becoming part of the official party machine.

This decision put real distance between Obama and his crowd, and limited its capacity to act to support his agenda; for example, it could not put pressure on those Democrats who were on the fence about key parts of the president's agenda, like his health-care law. It also alienated the many political independents, Republicans, and far-left types who had connected with Obama, but had no interest in being part of the formal political infrastructure of the Democratic Party.

Obama, a new president facing a big economic crisis, was intensely focused on enacting his policy agenda and, perhaps awed by all the tools of state now at his disposal, slipped into old power mode.

He tried to mobilize his supporter base around major moments in the making of his health-care law, and with some success, but he'd already moved out of a movement mindset—and it showed. OFA asked supporters to make generic "pledges" on various issues (such as support for a public option in the health-care bill, which Obama quickly dropped), but it was too cautious and had not tended to the community carefully enough to be able to whip up any real passion. Of course, as the Obama crowd was suffering from neglect, another well-funded popular movement was emerging to oppose it: the Tea Party, which mobilized mass protests and angry town halls to oppose health-care reform, presaging its huge victory in the 2010 mid-term elections, which crushed Obama's governing majority.

Obama's supporters wanted to do more than pledge. Had Obama invested heavily in enabling his supporters to launch local efforts— giving them freedom about how to organize in support of him—he might have built a strong, locally grounded progressive movement. This could have been a powerful counterforce to the rise of the Tea Party, especially outside of the major cities where Democratic support tends to cluster.

Obama's team eventually seemed to view his crowd more like an ATM machine than a movement. When the time came to run again in 2012, "big data" had replaced movement energy as the most sacred ingredient in the Obama tactical repertoire. The heroes of the re-election campaign were no longer the volunteer organizers (though he still had a good number of those), they were the backroom data whizzes. Clickbaity emails flooded our inboxes with subject lines like "Dinner?"—giving the recipient a momentary flutter of hope that the president was finally asking if we wanted to hang out. But Obama was also suddenly the kind of friend who always seemed to need to borrow money. The change he was waiting for often came in $5 increments.

Obama achieved a great deal without a hugely energized governing movement. He showed a respect for the office of the presidency and its conventions that his successor in large part has not. But once President Trump was elected and had the ability to sweep much of his legacy away, one was left wondering what might have been had Obama the president acted more like a genuine Crowd Leader—and less like a Cheerleader.

SIGNAL, STRUCTURE, AND SHAPE: THE THREE KEY NEW POWER LEADERSHIP CAPABILITIES

Today's leaders must succeed in a world not just of institutions, but of crowds. As we've seen, leading (and staying) in the Crowd Leader quadrant is a complex task. It is one that demands a particular set of skills. And an unlikely new power leader—Pope Francis—can teach us a lot about what is required.

Pope Francis's first act as the Bishop of Rome was to pray. At the moment he was elected, in fact, he was already praying, overcome by what he has described as a "great sense of inner peace," one he says he has felt ever since.

But his next three acts were, each in its own way, remarkable. He sent back the fancy red cape with the ermine collar that newly elected popes typically wear, electing to don simple white robes instead. He reportedly told the Vatican master of ceremonies, "You put it on instead. Carnival time is over!" He then broke with protocol as he greeted his cardinals, refusing the elevated throne, saying simply, "I'll stay down here." Finally, as he was revealed to the world on the balcony of St. Peter's, he did not ask for a blessing from his cardinals, nor did he ask for the success of his church. Neither did he, as was the tradition, offer his first papal blessing to the people. Instead he asked his people to pray so that "God may bless me through you."

It is hard to think of a role that is richer in symbolism than the papacy. And in his first hours as pope, Francis sent strong and resonant signals about how he was thinking about power, as much by what he did not do as by what he did. Gone was the super-hero cape. Out went the throne that raised him above his flock. That evening, he abandoned the papal limousine and took the minibus with his cardinals to supper. On his first night as pope, he slept in the guesthouse apartments, leaving the papal penthouse palace empty (a living arrangement that became permanent).

A Latin American cardinal with a track record of unconventionality, Francis had not expected to become pope. He had other plans afoot, with a place lined up in a retirement home for priests in Buenos Aires. The bookies didn't think much of his chances, either, staking him as a rank outsider at 33 to 1 in the pre-Conclave betting markets. So he had arrived in Rome anticipating a quick trip with just a small suitcase.

The church he now headed was badly damaged. Corruption was rife, especially around the Vatican bank. And the church's sins weren't just financial. Revelations of decades of child abuse throughout the church that had harmed thousands of its most vulnerable charges deeply damaged its reputation and had a direct impact on church attendance. Catholics' opinions of Pope Benedict had been relatively low and were falling in the later years of his reign.

Beyond these deep challenges, there was a general sense that the church was fading from relevance. Too few young men were joining the clergy in the United States. Many parishes were short of priests. In Rome, inefficiencies and inertia within the Vatican organization were all too clear. (This had been a long-standing issue. The famous joke among Vatican insiders is that when Pope John XXIII was asked how many people work at the Vatican, he answered, "About half.")

The church also found itself on the wrong side of major cultural shifts such as the growing acceptance of LGBT people. It was often drawn into public debates for what it stood *against*—abortion, contraception, women clergy—rather than what it stood *for*. On top of all this, Catholics had been rattled by the odd and almost unprecedented resignation of Benedict XVI. The church Francis inherited was, as long-standing pope-watcher BBC correspondent David Willey described to us, in a "terrible crisis."

In the story of how Francis has approached these big challenges, we can start to pick out the skills unique to leadership in a new power world.

Signaling is the way a new power leader makes a crowd feel more powerful through his speech, gestures, or actions. Obama's rhetoric of "we are the ones we have been waiting for" was classic signaling, designed to stoke his supporters' sense of agency and willingness to participate. The pope's request for, rather than bestowal of, prayers worked in the same way.

Structuring is how a new power leader puts in place structures and practices that enable the participation and agency she seeks to build. This is typically much harder work than signaling. Obama's 2008 campaign created a wide range of ways people could engage and not just feel ownership, but actually take it. (A variant of this is **unstructuring**, creating the broad space and energy for people to engage on their terms, but without any prescription or limitations. Donald Trump, as we have seen, intuitively grasps this.)

Shaping is how a new power leader sets the overall norms and direction of her crowd, especially in ways that go beyond her formal authority. When a leader is successful, these norms become so well understood that they are adopted and upheld by the crowd itself, and no longer rely on the leader. The ultimate aspiration of the octogenarian Pope Francis, as we will see, is to shift the norms of his church in ways that last well beyond his tenure.

Signaling: How Francis acts as "walking parable"

Within a very short time, the pope has become the world's most popular leader, with high favorability ratings across all faiths. He is even well regarded by atheists.

Much of this enthusiasm has been driven by his mastery of signaling, which he did so disarmingly at the time of his elevation to mark himself out as a different kind of pope, and the use of which has come to define his tenure. More than most previous popes, Francis has been encountered through a series of highly public and highly charged symbolic gestures, each of which affirm his values. Think of the time he traded use of the papal Mercedes-Benz for a well-loved Ford Focus. Or when he lingered in St. Peter's Square to bless and embrace a severely disfigured man. Or when he washed the feet of refugees. All these signals send a very clear message to his cardinals and his congregation as to how they ought to carry themselves in relationship to those less powerful.

In this sense, he has become a "walking parable," as Father Thomas Reese, senior analyst at the *National Catholic Reporter*, described it to us. The images and moments he generates have become modern icons—representations of the divine in the tradition of those once daubed on the cells of monks, or hung in the vestries of churches, but now going viral on Instagram.

Structuring for participation

Beyond his symbolic actions, Francis's tenure has been defined by an assertive reform effort, overhauling the structure of the Vatican bank, bringing greater transparency into the Vatican administration as a whole, and placing reformers into key positions. On one level this can be seen as a necessary reaction to corruption and calcification within the church. But it also speaks to his belief that the church should act as an "inverted pyramid," with the clergy ministering to the people, not living on clouds above them.

To realize this inversion, Francis has sought to push power away from Rome and toward local leaders and their communities.

Take the 2015 Synod on the Family. A synod is typically a gathering of bishops behind closed doors to discuss changes in doctrine and approach. This time, though, Francis mandated that questionnaires first be distributed throughout the Catholic world with the goal of "engaging families themselves, listening to their joys and their hopes, their sorrows and their anguish." "To my knowledge," said Dr. Thomas Groome, a professor of theology at Boston College, "it's the first time in the history of the magisterium [that they] have genuinely attempted to consult the laity."

Another initiative that pushed power down, this time away from the cardinals and toward local bishops and priests, was the groundbreaking—and headline-making—*Amoris Laetitia,* a papal "exhortation" on love in the family, where Francis suggested that local bishops and priests might make their own judgments on how to treat divorced and civilly remarried Catholics, including potentially allowing them to receive communion, something that the church had always forbidden.

By empowering parishioners, priests, and bishops, Francis is structuring for participation, creating ways for them to lead their own local efforts.

Shaping: The quality (and strategy) of mercy

"Who am I to judge?"

These are the most famous five words of Francis's papacy, spoken in reply to journalists on a flight back from Brazil, as he discussed the church's position on homosexuality.

These words, and their canonical implications, have been parsed in great detail. But looking for practical takeaways may be missing the point. It is through such statements that Francis sets direction for the church *without* actually creating new doctrine. (Note that this has been a criticism of him, too.) Shaping the norms of his flock is a subtler task than the traditional exercise of authority.

As mentioned, the pope has worked hard to steer his church away from the hot-button issues that have tended to divide and define it. He has been outspoken in attacking the clergy's "obsession" with a "disjointed multitude of doctrines to be imposed insistently." Francis wants his church to expend its energy (and capture the public's imagination) through its core work of serving the poor and becoming a "home for all." His remarks on refugees in the early weeks of President Trump's administration point clearly in this direction: "It's hypocrisy to call yourself a Christian and chase away a refugee or someone seeking help, someone who is hungry or thirsty, toss out someone who is in need of my help."

He has done this, too, through an emphasis on *mercy,* perhaps his defining theme. His words "Who am I to judge?" don't change the church's letter of the law on homosexuality, but they do effect a real shift in focus. He is attempting to transport his church away from an old power paradigm, defined by clergy handing down judgments on their people, chastising their behaviors, dividing them into saints and sinners, the included and excluded. He is gradually shaping a church that focuses less on inward debates about rules and more on outward demonstrations of its core values. As he has noted, "Mercy is doctrine." By shaping new norms, and by enlisting the church's millions of followers and not just its few leaders in doing so, the rules themselves may follow.

Yet the deep moral challenges posed by the church's child sex abuse scandals aren't going away, and Francis is not confronting them with the same zeal he has brought to other issues. Take the case of Australia's Cardinal George Pell, handpicked by Francis to lead financial and other reform efforts in the Vatican even as he was being roundly criticized for his lax handling of child sex abuse cases over several decades. In 2017, Pell himself was charged by Australian authorities with child abuse offenses and was finally forced to take leave.

If the pope does not confront all forms of corruption at the heart of his church, and instead seems to put protection of his favored clergy above respect for the laity, as his predecessors have too often done, then his efforts to bring new power to the church will ring hollow. The biggest tests for Francis may be yet to come.

np = respect?

THREE SKETCHES OF NEW POWER LEADERS

Francis aside, leaders aren't very popular these days. The 2017 Edelman Trust Barometer shows that only 37 percent of people think of CEOs as credible. And trust in all four institutions tracked—business, government, NGOs, and media—to do what is right is in decline, the first time Edelman has observed this since they began measuring trust.

A range of factors—the rise of new power forces among them—has left a generation of leaders relying on behaviors and ideas that suddenly feel out-of-date, in many cases after serving them well for decades. And this is not a problem that can be solved with a simple emotional pivot. The conventional wisdom that encourages our masters of the universe to shapeshift into masters of empathy—aligning themselves with the twin moons of humility and authenticity—does not go far enough. In fact, it can distract leaders from the more difficult task of moving the focus away from them and onto their communities, a skill mastered by the three leaders profiled below.

How to be a Genius in the new power world: The wisdom of Ai-jen Poo

When Ai-jen Poo won the MacArthur "Genius" Fellowship, she went well beyond the default line of acceptance: "I couldn't have done this without my colleagues. This is their award too."

She and her team at the National Domestic Workers Alliance (NDWA) invested her $625,000 MacArthur grant—intended to facilitate Poo's own creative growth—into the growth of the alliance's workers instead. Thus was born the Dorothy Lee Bolden Fellowship, which would create leadership development opportunities for the group's vast community of nannies, housekeepers, and caregivers.

Bolden was a key early figure in advocacy for domestic workers. She founded the National Domestic Workers Union in 1968, paving the way for Poo and her colleagues, who founded the National Domestic Workers Alliance, which gives voice to 2.5 million domestic care workers in the United States.

Poo's use of the MacArthur stipend sent a potent message: *This is about us.* By honoring a woman who came before her and marking her faith in the future of the community she serves, she positioned her prize as the validation of a *movement*.

This moment was a testament to Poo's character—when you meet her, it's clear she's doing this work for all the right reasons. But it was also a strategic imperative. Poo's effectiveness as a leader relies on striking the right balance between her prominence and her crowd, a group of marginalized, mainly immigrant and black women who are among society's least powerful.

As she put it, her job is to use her power in a way that "creates power for more people." This is a special challenge in her case, because a severe imbalance in power is something her constituency experiences every day: "They really understand inequality because they live in poverty and then they go to work in the most extreme forms of wealth." Beyond the wage and wealth gap domestic workers experience viscerally, they sometimes tell horrific stories, too: of neglect, abuse, and threats.

For Poo as a leader, managing this tension has not always been easy. When she was in her twenties, there were fierce debates as to whether she, who had never been a domestic worker, could legitimately represent the interests of domestic workers and lead their movement. Ironically, the better she did in galvanizing what had previously been a set of disparate groups and a largely neglected community, the more she found herself portrayed as someone who lacked the "street cred" to lead.

She has managed this tension, and held her ground, with an understated and collaborative style, taking care that others are seen to take on important assignments and ensuring that her governing structure includes those she serves. The majority of her board members (whom she refers to as her "boss") aren't policy experts or donors, as they are at most nonprofits: they are domestic workers. She estimates their average annual income at $22,000.

Every two years, Poo brings together the NDWA community to elect the board and set the alliance's course. Five hundred domestic workers come together for debates on their organization's key issues and challenges, from salaries to strategies. Proceedings are interpreted in eight languages so that everyone can take part.

Those who want to run for the board get the chance to step up and address the assembly. It is a tense scene. Each candidate gets three minutes to speak. Some thrive. Others fumble. All get cheered on. There's a candidate debate over the key questions and challenges facing the workforce. Then everyone gets to vote. Poo thinks of this process as one of the most meaningful aspects of her work at NDWA. As she told us: "Everyone was on the edge of their seats because they owned the organization, you know?"

She has structured the NDWA to make domestic workers protagonists, not mere beneficiaries. As she puts it, "The whole organization is basically set up as a platform to give voice to that constituency of people who are completely isolated." She thinks of everyone in her world as an organizer. And these are not just words. NDWA's Strategy, Organizing, and Leadership initiative was developed to train leaders who in turn work to develop the leadership and political skills of domestic workers. That initiative has strengthened the capacity of groups across the network, with an estimated 3,000 workers learning to build movements and advocate for their rights.

Poo's unique position demands that she be extra-conscious of issues around status and signal in a way that shows she is truly investing in her crowd. To lead her crowd effectively, she has to constantly demonstrate how she can blend into it.

From Neutron Jack to Electron Beth: A new style of leader at GE

From time to time, Beth Comstock will present one of her colleagues with an interactive Egg Minder. This novel device uses Wi-Fi to connect you with the eggs in your fridge, so never again could you be at the supermarket and find yourself unaware of how many eggs you had left or how old they were. It's the kind of device that delights technophiles but most others find faintly ridiculous. The product was an epic flop.

Handing out those surplus Egg Minders lets Comstock, GE's vice chair, say something about the culture of innovation at GE. The Egg Minder was part of GE's partnership with Quirky, a much-hyped new power start-up that allowed the crowd to invent their own products. Comstock herself had championed the Quirky relationship and had been closely tied to it when the company stumbled and then filed for bankruptcy.

The signal she is trying to send, therefore, is clear: *I failed at this and I'm still here.* It's not just OK to engage the crowd and fail spectacularly at GE; it is expected and embraced.

Consider how different a philosophy this is from that of the legendary GE leader Jack Welch, who made his reputation through elimination of error and failure with his Six Sigma management process, and through elimination of staff with ruthless and regular culls. It was the layoffs that earned him the nickname "Neutron Jack," because he'd explode and empty a building of people while leaving the structure standing.

Comstock leads very differently. In GE nickname terms, we could think of her as Electron Beth, defined not by how she blows things up, but by how she binds people together. Her success in leading GE into the digital age has earned her promotion up the ranks and caused her to be named one of the 100 most powerful women in the world.

At the heart of her success is how she has structured GE to deliver greater creativity, inside and out. She targeted not just the 300,000 people on GE's payroll, but the vast communities beyond its walls whose involvement she sees as key to her company's future.

The Quirky story was part of an effort led by Comstock around open innovation, the same approach to crowd-based problem solving tried out by NASA. She chalked up early wins like the Engine Bracket Challenge, where GE partnered with GrabCAD, a digital platform of more than a million designers and engineers, to dream up a lighter, more efficient part for its planes. The winner was a young Indonesian engineer with no aviation experience whose idea delivered an 84 percent improvement.

She built on these successes through partnerships with organizations like Local Motors, the first-of-its kind new power car company we'll profile in the next chapter, to crowd-source product development challenges to engineers, coders, and scientists, both online and in microfactories across the United States.

Within GE, Comstock called on Eric Ries, author of the popular book *The Lean Startup,* to consult on a new way to encourage and speed up the company's internal innovation and product design efforts and quickly incorporate early customer feedback. This led to the creation of the FastWorks program, which has now trained over 40,000 of GE's leaders.

FastWorks is leading a shift toward experimentation and prototyping, nudging norms at a company whose stock price has been in long-term decline and that some say has become too big to innovate. It has also led to some big wins (as well as, we can safely assume, a couple of new Egg Minders). A project to create a digital wind farm delivered a solution in less than four months and delivered a 20 percent increase in renewable energy for each wind farm. GE has estimated that FastWorks will save several million dollars per project in product development costs.

Comstock is trying to push power down, to "distribute the decision-making process as widely across the network as possible. In effect, to empower individual 'cells' to relay signals and respond to their local conditions as they see fit," as she puts it. This has meant redistributing people to staff up local markets around the world, as well as attempting to shift cultural practices like feedback. In place of the annual performance review, experienced in most organizations like a dreaded annual dental checkup, GE now uses mobile tools that allow many more people to offer insights and highlight issues in real time. We'll discuss the importance of feedback in the new power workplace in chapter 11.

Peter Sims, the entrepreneur and writer, has known Comstock for years and seen her up close as she has risen in the ranks. He describes how she flows freely among communities of artists, makers, and robotics wizards—rather than just spending her time with the suits. As Sims put it, "While she has a very big job, and a very big title, nobody who interacts with her experiences that."

By sourcing new ideas from unexpected places, Comstock is blurring the line between the crowd and the corporation.

How Lady Gaga unleashed her monsters

Die-hard Barry Manilow fans are known as Fanilows (you know who you are). Justin Bieber has his Beliebers. Demi Lovato? Lovatics. Ed Sheeran? Sheerios. (Really.) But Lady Gaga's fan base has embraced a different kind of identity.

They are known as Little Monsters.

The name had emerged early on in Gaga's relationship with her fans, but after she saw the movie *The Social Network,* about the rise of Facebook, she set out to create her own social network to bring her closer to her fans and her fans closer to one another.

Her vision was of much more than a Lady Gaga fan page. She wanted the site and its branding to be not primarily about her work, but about fostering her fans' confidence and artistry. The result, LittleMonsters.com, allowed her legions to share their comments, art, ideas, and stories with one another, as well as to organize events. She quickly had a million participants.

For Gaga, this was a mission that mattered: many of her fans were girls confronting brutal body-image expectations and sexism, and LGBTQ youth grappling with questions of identity and difference. The big payoff for connecting her fans was that it gave her a tool to promote values they could embody well beyond their interest in her.

Monster @HausofFanis put it like this: "Our bond is beyond just loving the same star. Our bond is based on the philosophy we share. We all believe in the importance of love, compassion and tolerance." Hector, a seventeen-year-old Colombian teenager, defines a Little Monster as someone who "doesn't judge or criticize what others are, regardless of their sexuality or skin color or other traits and stuff."

It should perhaps be no surprise that Gaga, who often experiments with her own identity, has chosen to encourage her community's confidence in their own. As Professor Amber L. Davisson writes, "Rather than Gaga pushing her fans to adopt a specific identity, the narrative of the little monster is about embracing their own identity."

In reclaiming the word "monster," they have turned difference and marginalization into strength. The Little Monsters community offers a textbook example of "optimal distinctiveness." Monsters feel just the right amount of the same and just the right amount of different.

The spirit of LittleMonsters.com infuses Gaga's entire relationship with her fans. Her music itself speaks clearly to these themes ("Born This Way" having become something of a national anthem for the monsters), and so does her dialogue with monsters via her huge social media streams, where she regularly shares the art and ideas of her fans. She is a champion of *their* creativity. She describes herself (and is referred to as) "Mama Monster," more matriarch than superstar.

In doing all this, Gaga has over many years not just built a more die-hard fan base for when her albums drop, she has developed a broad constituency she is able to steer at key moments. In 2010, Gaga—escorted by four gay service members, two men and two women—presented at the MTV Video Music Awards, asking viewers to call their senators to get them to repeal the "Don't Ask, Don't Tell" bill. She held rallies in Portland, Maine, asking fans to call their senators and tweeted Senator Harry Reid, then the Senate majority leader, to push the vote. With Gaga's call to action, senators' lines were flooded with so many phone calls that the switchboard was overloaded. Her campaign forced the Senate to update their phone system to better handle incoming citizen calls. As one headline put it: "U.S. Senate Getting Gaga-Compliant Phone System."

—

It is no coincidence that all three examples of new power leadership above are women. In her analysis of the research addressing differences in the leadership styles of men and women, Alice H. Eagly observes that "female leaders, on the average, are more democratic and participative than their male counterparts. Men, more than women, adopt a top-down, 'command and control' style."

Eagly's findings speak to a much earlier insight by Mary Parker Follett, a pioneering management thinker. In her 1918 book, *The New State,* she sketched out a fresh view of how to lead that played down "power-over" in favor of "power-with, a jointly developed power, a co-active, not a coercive power." This idea presages an important strand in contemporary feminist thinking that emphasizes more participatory and equitable approaches to power. Today, enabled by a proliferation of technologies that supercharge collaboration, Follett's vision feels more attainable—and more necessary—than ever.

Of the three new power leadership skills on display in these stories, perhaps the most difficult to master is shaping values and behaviors beyond the bounds of one's direct control. For Gaga, that meant kindness, diversity, and tolerance. Comstock championed a new culture of experimentation. Poo fostered pride, dignity, and the willingness to confront inequality rather than accept it as inevitable.

Not everyone succeeds. Next we'll turn to a contrast between two approaches to leadership on a difficult issue: race. One approach was very new power, the other very old, and they led to *very* different outcomes.

#RACETOGETHER VS. #BLACKLIVESMATTER: LEADERSHIP IN A TIME OF HASHTAGS

"It began with one voice," read the opening lines of Starbucks' press advisory announcing its "Race Together" campaign. It continued: "As racially-charged tragedies unfolded in communities across the country, the chairman and CEO of Starbucks didn't remain a silent bystander. Howard Schultz voiced his concerns with partners (employees) in the company's Seattle headquarters and started a discussion about race in America."

To explain his vision for Race Together, Schultz posted a six-minute video featuring himself talking over footage of racially diverse Starbucks employees participating in company forums about the initiative. "The last few weeks I've felt a burden of personal responsibility not about the company but what is going on in America," he declared.

He then encouraged his baristas to write the hashtag #RaceTogether on customers' coffee cups and engage them in conversation about race in America. Schultz, a white baby boomer billionaire, was rallying his army of young low-wage workers, a substantial percentage of whom are not white, to make time during the forty-five seconds or so that the milk was being steamed for a double-chai-fat-free-soy-latte to stage what would become America's most awkward teach-in.

The reaction was swift and harsh. As @IjeomaOluo pointed out: "Being a barista is hard enough. Having to talk #RaceTogether with a woman in Lululemon pants while pouring pumpkin spice is just cruel." Or as @ReignofApril put it: "Not sure what @Starbucks was thinking. I don't have time to explain 400 years of oppression to you & still make my train." The backlash was such that Starbucks' head communications honcho ran for cover and temporarily shut down his Twitter handle. The story took over the Starbucks' annual meeting, held just after the campaign was launched, overshadowing the company's impressive results. A couple of weeks later, Starbucks ended the campaign "as planned." But it was clear that it had not gone at all as planned.

Schultz raised an important subject that deserved national attention at a time when it would have been safer just to launch another warm-and-fuzzy campaign on recycling. But his initiative is a fascinating case of how even a leader with good intentions could have gotten the three Ss—signaling, structuring, and shaping—so wrong.

"We made a tactical mistake. So what?" Schultz told *Fast Company* about Race Together. "We're moving forward."

But this was more than a tactical mistake.

First, the campaign, beginning with its six-minute video, was all about Howard. He was presented as the hero leader bravely raising an issue other CEOs wouldn't touch. He then partnered with another white billionaire, Larry Kramer, the publisher of *USA Today*, to place an op-ed (with a picture of Schultz at the top) to promote the campaign.

This approach crowded out the people who might have helped him make the initiative work. By taking up so much space himself, there was no room left for others to occupy.

Race Together seemed hasty and ill thought out, but this was no whimsical move. Schultz had tested and developed his thinking with a fair amount of consultation. He had been encouraged by the enthusiasm he had seen at a range of "open forums" among his corporate staff. Yet for all this engagement, the flaws in his plan had not surfaced.

This is one of the big challenges of the "listening tour," ever popular among CEOs. There aren't a lot of employees, even relatively senior ones, with the gumption to say, "Great to meet you, boss, love your passion, but you're dead wrong on this one." And his frontline workers, who were blindsided by the campaign, were left wondering how on earth this had happened in the first place: "If we're going to have a discussion about race, how ethnically diverse was the group of people who decided this was wise?" one barista asked at the time on Instagram.

Simply "consulting" your base is a rudimentary application of new power and not likely to generate meaningful feedback. Truly structuring for participation requires more. Paradoxically, in the absence of this, an old power strategy might have helped: holding closed, online, or confidential forums can create a pathway for brutally honest feedback.

It is easy to mock #RaceTogether (and many did). But there was something to the idea. Schultz understood that coffee chains like Starbucks have become a de facto twenty-first-century community center (albeit one that favors gentrified neighborhoods). Starbucks stores around the country might have been used, perhaps alongside other local spaces in poorer areas, to host local events moderated by organizations with experience facilitating difficult conversations about race. You can imagine, too, how Schultz might have recruited his diverse army of baristas not to watch *his* video and write *his* slogan on paper cups, but to build something of their own.

#RaceTogether started with just one voice. It would finish that way, too.

Schultz's attempt to exercise leadership on race issues did not come out of nowhere. Race Together emerged in the context of a national movement led by activists who had very little in common with Schultz and who have adopted a radically different leadership approach: "3 black women started #BlackLivesMatter. 1 is Nigerian-American, 2 are queer. Complicate the narrative . . . Because all #BlackLivesMatter," declared the official @blklivesmatter Twitter account, several months after activists and others around the United States had taken up the slogan.

Black Lives Matter is a meme that became a movement and has shifted the conversation about race in the United States. The activism it inspired has forced changes to policing practices, boosted black turnout in some elections, shaped legislation, booted out police chiefs, and drawn renewed political and cultural attention to the systemic injustices black Americans face.

It began with a Facebook post.

In 2013, after George Zimmerman was acquitted of the killing of the black teenager Trayvon Martin, activist and organizer Alicia Garza wrote a powerful post that ended with the refrain "Our Lives Matter, Black Lives Matter." Garza's friend Patrisse Cullors turned this into the hashtag that became a global trend and an umbrella for fighting injustice, #BlackLivesMatter. Opal Tometi started to set up social media accounts to broaden their campaign.

It soon began to catch on, and people started to offer their help in building the movement. "Cultural workers, artists, designers and techies" all offered their talents. The founders' tactics became a model example of the crowd-building approach we discussed in chapter 4. Garza, Cullors, and Tometi set about, in Garza's words, "[creating] the infrastructure for this movement project—moving the hashtag from social media to the streets." They began to mobilize online and on calls, connecting organizers around the country, with the goal of creating the "space for the celebration and humanization of Black lives."

Their response to the killing of Michael Brown by police in Ferguson, Missouri, took the movement to the next level. The three women—working as always in partnership with others—quickly organized the Black Lives Matter "Freedom ride," where activists from around the country piled into buses and rode to Ferguson to support local organizing efforts and the community there.

This iconic moment inspired many more. Three months later, in December 2014, a Black Lives Matter banner stretched across the crowd at the Millions March NYC. This 50,000-strong march was initiated by twenty-three-year-old Synead Nichols and nineteen-year-old Umaara Elliott, who had never organized a protest before.

From its start in a Facebook post, Black Lives Matter exploded across the country, grabbing national headlines and growing into a cluster of organizations, local chapters, and loose-knit swarms of organizers. Its decentralized design and "ownerless" identity allowed it to be owned by many and become influential over millions.

So what did #BlackLivesMatter get right that Schultz got so wrong?

Obviously, the #BlackLivesMatter founders had an inherent legitimacy and credibility in addressing race issues that Schultz did not. But there is more to it than that.

From the start, the movement's signals were all about collective participation—de-emphasizing the role of any one individual leader. Alicia Garza told us that this was not about making Black Lives Matter leaderless, but rather leader-*full*. "When we think about what happened to Martin Luther King, or what happened to Medgar Evers, or Malcolm X, I mean all of these men were assassinated because of their leadership role," Garza told us. "Frankly, their organizations were not the same afterwards. The reason for that is because even the organizations themselves kind of built themselves around a person. . . . When you cut off the head, then the rest of the body dies." In her work shaping Black Lives Matter, she is deeply cautious about "isolating or making exceptional a few people when, really, the piece that's exceptional is that everyday people are taking this up and trying to make it into something much bigger than all of us."

Over time, those many leaders have built scaffolding to structure and support the movement, too. Yes, there is a lot of free-form activity around the movement (a good example of unstructuring). But the co-founders and others have also carefully developed local chapters. As Jelani Cobb describes in *The New Yorker*: "Prospective chapters must submit to a rigorous assessment, by a coordinator, of the kinds of activism that members have previously engaged in, and they must commit to the organization's guiding principles."

But what is most notable, and important, about Black Lives Matter is its commitment to radical inclusivity. This is a movement that wants to have lots of leaders, and it is committed to shifting the spotlight onto those who had not typically been cast in that role, and who have been the most marginalized. "The other piece of this involvement is: how does Black Lives Matter *really* push the narrative that *all* Black lives matter? . . . So the nature of the work is about shaping, trying to shape this network politically," Cullors says. As the women lay out in the "About Us" section of their website: "Black Lives Matter affirms the lives of Black queer and trans folks, disabled folks, Black-undocumented folks, folks with records, women and all Black lives along the gender spectrum. It centers those that have been marginalized within Black liberation movements."

Interestingly, Cullors, Garza, and Tometi have repeatedly staked their claim as co-founders of Black Lives Matter in part because they aren't straight men, even though others in the movement have criticized them for trying to take credit. The three are adamant that they do not want to replicate the old power model, with a charismatic speechifying male preacher like Al Sharpton or Jesse Jackson taking up all the space. This led to clashes in the early days of Black Lives Matter. Jackson found himself booed when he tried to deliver a speech to protesters in Ferguson. Sharpton was accused of trying to co-opt the movement. At the Justice for All March in Washington, D.C., in late 2014, Sharpton and his group literally cut the mic and silenced a young activist, Johnetta Elzie, who had been tear-gassed at the Ferguson protests, preventing her from speaking. She and her friend were told they couldn't share their views because they didn't have a "VIP pass," she told *The Root*. "If it is a protest," she wondered, "why do you need to have a VIP pass?"

It is striking that perhaps the most recognizable figure to have emerged in the Black Lives Matter era is a man, DeRay Mckesson, a journalist and activist who was part of the early Ferguson protests and has a familiar charismatic leadership style. Politicians and the mainstream media have gravitated to him. Garza wonders whether "the phenomenon around DeRay is really about people feeling more comfortable with men in some form of leadership role." A challenge for leader-*full* and women-led movements will be getting the old power structure to see and engage with them for what they are, rather than trying to turn themselves into something those existing structures can recognize.

Black Lives Matter offers a case study in new power leadership. The three founders have set a course that has already outlived their own roles. Some of the many leaders who have emerged from the movement, like Mckesson, have run for office, reflecting the variety of views within the movement about how to make change happen. As in any broad-based movement, many factions and disagreements have emerged. But the achievement is clear. Black Lives Matter placed a much-neglected issue at the heart of the national conversation. As they look ahead, the next task is to translate all the energy they have created into lasting change, both political and cultural.

CONCLUSIONS: REVISITING THE LEADERSHIP COMPASS

Let's now populate the compass with the leaders we have featured in this chapter.

There are many ways to be a **Crowd Leader**. Some, like the Black Lives Matter founders, are radical in their embrace of both new power values and tools. Beth Comstock, in contrast, operates

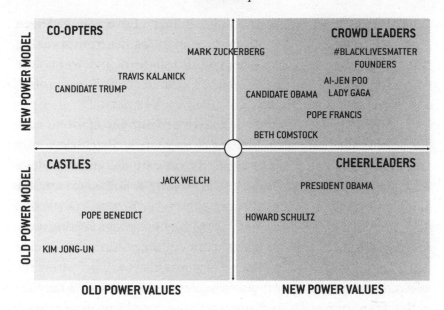

within a traditional corporate context and has structured for participation inside and outside of GE in more bounded ways. Style is a factor, too. New power leaders like Comstock, Poo, and the Black Lives Matter founders have quite assiduously avoided becoming charismatic leaders while others use their personal magnetism strategically in order to pursue their goals, like Obama and Pope Francis. (This is a universal fact of leadership—you can see examples of both charismatic and more understated leadership styles in every quadrant.)

The worlds of politics and business today, particularly in the United States and Europe, are chock full of **Cheerleaders** like Howard Schultz and President Obama—who embrace new power values but have work to do in deploying new power models. Other politicians who built a movement to get elected, like French president Emanuel Macron, face similar choices and dilemmas as they attempt to translate movements to government. Some here are earnestly trying to work all this out, gradually heading up into the Crowd Leader quadrant. But others get a little too comfortable, essentially operating as before, but now associating themselves with the crowd in well-publicized but insubstantial ways.

We haven't spent much time talking about **Castles** because this is the model of traditional leadership we're all overfamiliar with. Still, it's worth noting that this kind of top-down leadership can take many forms—from unabashed dictators like Kim Jong-un, who uses traditional propaganda techniques paired with extensive state violence and coercion to achieve his aims (even if this means killing his relatives), to much more conventional business and political figures like "Neutron Jack" Welch or Pope Francis's rigid predecessor, Benedict. What's striking about the strongman model of leadership is how, in a world of increasingly effective crowds, this model has come to feel limited. Why be a regular old strongman—relying purely on your formal authority and capacity to enforce the rules—when you can be a Platform Strongman, wielding your very own crowd but still keeping power concentrated in your hands?

Which brings us to the **Co-opters**. The Co-opters are powerful because they are proficient in the new ways of participation. They can engage and delight and inspire and direct (or misdirect) a crowd so effectively that the crowd might be too busy or distracted to even notice that it is not actually getting more powerful. Consider Mark Zuckerberg. As a leader, he's the master of building a crowd—he sits at the helm of what is, by many measures, the world's largest. Zuckerberg says he has renewed Facebook's mission to focus on global community building and he seems sincerely committed to this goal. But will Zuckerberg pursue this mission if it means giving up some of his own power? Sharing more economic value with users, giving his community a real say in Facebook's governance, and giving them the ability to see, shape, and tailor its algorithms appear to be off the table. Instead, he has positioned himself as a benevolent steward who can be trusted to make decisions in the best interests of the network. Zuckerberg straddles the Co-opter and Crowd Leader quadrants because, as we write this, he stands at a crossroads.

The future is going to see a leadership contest between Co-opters (and the Platform Strongmen they often become) and Crowd Leaders. At their best, both can signal, structure, and shape highly effectively. But these two archetypes tend to produce very different outcomes in terms of how power is distributed.

In many ways, it is the Crowd Leaders whose work is hardest. Their proposition is more complex than "I alone can fix it." They are asking people to do more, to take on greater responsibility, and to collaborate more freely.

And even if that is the kind of world we might all want to live in, it's far from clear it's the one we'll get.

THE ART OF BLENDING POWER

A big theme of this book has been the contrast and conflict between old and new power. Think back to the definitions we shared right at the beginning, expressed in the chart below.

Old Power	New Power
Currency	Current
Held by few	Made by many
Download	Upload
Leader-driven	Peer-driven
Closed	Open

Many of the stories in the book speak directly to these oppositions. Look at the way the Ice Bucket Challenge did away with the telethon. Recall the friction between the solution seekers and the problem solvers at NASA, or the reform-minded pope and the intransigent Vatican bureaucracy.

This chapter is not about the tension between new and old power. It offers stories about organizations that have worked out how to blend power, creating models that bring together old and new power in ways that reinforce each other. Instead of seeing the chart's pairings as dichotomies—open or closed, download or upload—these organizations have found that the combination of old and new power can be greater than the sum of the parts. An essential theme of all their stories is how they know when to dial each one up.

WHY THE NRA KEEPS ON WINNING

Leading with new power: The Colorado recall

In September 2013 politicians all over the United States were sent a very clear message. Two Colorado state senators who had helped to pass laws aimed at limiting gun violence were unceremoniously removed from office in a special recall election held to target them for their stance on guns. This victory for gun rights advocates was remarkable coming nine months after a massacre in a Newtown, Connecticut elementary school had left twenty-six dead, most of them small children, and in a state still haunted by the 1999 killing spree at Columbine High School.

Just two months after the recall election, a third Colorado state senator, Evie Hudak, preemptively resigned, simply because she knew she had stepped into the same crosshairs.

The Colorado recall election is an example of the remarkable power of the National Rifle Association (NRA) and its allies to foil those who cross them, and thus send a warning sign to others tempted to do so. Even when public opinion is overwhelmingly in favor of reasonable regulations (92 percent of Americans—including 87 percent of Republicans—favor background checks for all gun buyers), the NRA has time and again managed to restrain the forces of gun control. It goes on the offensive and comes out on top, even when it appears to be at its most vulnerable.

Why does the NRA keep on winning? You might assume that

its clout comes from the brute force of its well-funded lobbying muscle. After all, the NRA has a $250 million annual budget and a reputation for buying elections.

Yet there is more to it than that. The two state senators who faced recall were being defended by the billionaire mayor of New York, Michael Bloomberg, the country's strongest advocate for increased gun regulations. Bloomberg easily outspent the NRA and its allies, plunging over $3 million into the recall fight, while the gun rights crowd invested just $600,000. And not only was the NRA not spending as much, it also appeared to be somewhat missing in action during the standoff, not obviously out front, much to the consternation of some of the state's gun rights advocates in the lead-up to the vote.

Although they may not have been conspicuous, the efforts of the NRA were central to the Colorado recall. Over the years, it had made small grants totaling hundreds of thousands of dollars to dozens of organizations in Colorado—gun clubs, shooting ranges, conservationist and hunting organizations like Ducks Unlimited, and small local gun rights organizations not formally part of the NRA. These were investments designed to create a web of grassroots activity fusing together the politics, culture, and commerce of guns. Think of it as an independent, self-organized crowd, untethered officially yet strategically aligned with the NRA's interests.

Out of this community came activists like Victor Head, a twenty-eight-year-old plumber who, along with his friends and family, started the effort to gather the thousands of signatures required to trigger a recall election, after the idea had bubbled up in the dense world of online message boards for gun rights supporters. "The reason my brother and I got involved is that these laws really affected us," Head told a local TV affiliate, Fox 31 Denver. "That's where our passion came from. We used volunteers only who actually cared. They weren't canvassing for a paycheck."

Mark Glaze, a seasoned D.C. political operative and lawyer who at that time ran Bloomberg's gun control group, was in charge of orchestrating resistance to the recall effort. He found his group out-maneuvered.

He told us why the NRA's tactics proved so successful. "The benefit of pushing money downward and letting people do with it what they will is that very often they will do with it what you want. So when the recall effort came the NRA appeared to be virtually a nonplayer . . . Nevertheless, those grantees were so well-funded and so empowered that they took up electoral arms against these legislators and managed to take them out."

This was a world away from Bloomberg's strategy, which focused on lots of outside spending, and exerted much stricter centralized control over their grassroots: on his side, said Glaze, "we will pay organizers and make grants to local chapters but it all is on a very narrow and controlled basis."

Looking back, Glaze concedes that what the NRA did so well was "getting off your own agenda for five minutes." By releasing control of its crowd, at the potential cost of being able to claim a win, it created the freedom that gave people true ownership of the movement. (We'll say it one last time: It's only a movement if it moves without you.) Those who criticized the NRA for being AWOL during the recall campaign had misread the strategy. It had not disappeared. It had simply prioritized its ultimate mission over its brand.

While Bloomberg brought big funding, the NRA was able to draw on what proved a more valuable currency: intensity. Its early investment in a network of people like Victor Head meant it could count on a community on election day that was multifaceted yet singularly focused, turning the election into a referendum on guns. The NRA's job was to get out of the way, providing just enough air cover and institutional support to guarantee victory. As Glaze put it, "even though the NRA was largely absent they had sort of fed these passions over the course of years and their supporters were just much more pissed off about these gun laws than people on our side were delighted with them."

Leading with old power: The defeat of Manchin-Toomey

To win the Colorado recall, the NRA let its new power base do the work, and let its old power apparatus take a supporting role. But to beat the Manchin-Toomey bill—a gun control bill supported by President Obama and his allies in Washington—the NRA needed to flip its tactics, leading with its old power brand and then letting the crowd show up as cavalry.

At first, this bill looked like the kind of sensible gun reform that might finally pass. It, too, was offered up in the wake of Sandy Hook, and the overwhelming majority of Americans supported the basis of the proposed law, universal background checks. What's more, the bill's sponsors were two senators considered very strong NRA allies, not anti-gun bleeding hearts. Joe Manchin was a Democrat and a proud NRA member from the gun culture heartland of West Virginia. Pat Toomey was a Republican.

Yet the Manchin-Toomey bill was soundly defeated by the NRA, through the use of perhaps the most feared old power weapon in its arsenal. The NRA decided to "score" the bill.

The NRA's "score" is the letter grade it awards politicians, based on their voting record on gun rights. The scores run from A to F, and a lawmaker's grade determines how the NRA doles out carrots and sticks at election time. Before this bill, Toomey's score was a solid A minus.

As Doug Schoen, the pollster and analyst, explains in his detailed study of the way the NRA wields its power: "Tellingly, the NRA had declined to 'score' the earlier vote that brought the bill to the table in the first place, and that measure received 68 votes, well over the number it needed. However, after the NRA announced that they would, in fact, be scoring the vote on the key amendment on background checks, a full 14 Senators who had previously voted 'yes' to bringing the legislation to the Senate floor removed their support from the bill, and voted an NRA-friendly 'no' on background checks."

The NRA's grading system is the modern equivalent of the

emperor's thumb, pointed up for those he will save, and down for those he condemns. The halls and lobbies of Washington are full of stories of those who have crossed the NRA and paid a very high price.

Yet there is a twist. Schoen's study of multiple election cycles over the last fifteen years shows that the NRA's projection of power holds a large element of bluff. It often takes credit for races the pro-gun candidate was sure to win anyway, or where it invested minimally. In 2012, for example, the NRA claimed victory in five races in which it invested less than $100. Schoen shows that not only is the NRA not one of the biggest spenders in U.S. elections, as it is widely thought to be (it ranks only about 50th among outside groups in U.S. elections), it is not even especially effective in how it invests when it does so directly. The NRA extracts what we can think of as a *return on fear,* with its legend and threat looming larger than the empirical reality.

Yet that threat is made more potent and credible through the application of new power. When it came time to swing those votes in Congress, the NRA also used the intensity of its supporters to overwhelm the other side. Heidi Heitkamp, then a newly elected senator from North Dakota, recalled the pressure politicians were under: "I think I always had a reputation as somebody who will listen, somebody who is pretty independent-minded but also believes that at the end of the day, you got to listen to your constituents." She reported that, even though there was an extensive campaign in her state to get her to support the bill, when all was said and done, the calls her office received were "at least 7 to 1 against that bill."

Culture, commerce, and politics: How the NRA builds their intensity machine

Think about this number for a moment. Polls show that more than 90 percent of U.S. citizens favor legislation like that which Manchin-Toomey proposed, yet they were easily rolled over by the dedication and commitment of those advocating gun rights.

And where did all these calls come from?

Being a member of the NRA isn't just a political affiliation—it provides tangible benefits like discounts on insurance, access to personal loans, and a wine club. And it starts from childhood; the NRA's blog carries posts like "Marksmanship Programs Your Child Will Love." You can even buy a toaster that sears the NRA's brand into your bread. This is the NRA's "formal" crowd, members who have been engaged through the programming that has sustained it for decades. And it can also draw on its "informal" crowd, that messier ecosystem of people like Victor Head who make up gun culture and will rally when required.

The NRA is playing a long game. It is committed to building deep grassroots engagement and creating rewarding feedback loops that go far beyond its narrow political objectives. A close observer of the NRA, Jennifer Carlson, puts it this way: "The gun control side is a political movement. They are working to change gun policy. That's what they do. The gun rights side, they're changing culture, they're operating within an industry, and they are a political movement."

It is the power of gun culture that strengthens the NRA over time. As Mark Glaze notes, "By stoking nationalist claims and concerns about government tyranny, they prompt people to not only join up with a gun group, not only do the NRA's bidding and write the state legislature and vote against said state legislature, they're also buying more guns on the premise that the government is going to take their gun away." The NRA has created such a powerful unifying worldview that at moments of crisis, like mass shootings, it paradoxically gets stronger. More guns are bought, its members grow more fearful of confiscation and more committed, and a strangely virtuous cycle emerges.

And the NRA is rarely afraid to raise the stakes in its efforts to build up a unified worldview—and present a common enemy. In 2017, it issued a recruitment video that didn't even mention guns. In what many viewed as a barely disguised call to violence against political opponents of Donald Trump's presidency, and liberals in general, the narrator declared, "The only way we stop this, the only way we save our country and our freedom is to fight this violence of lies with a clenched fist of truth."

Lessons learned: The future of the gun fight

There's a lot we can learn from the NRA, much as many might wish otherwise. Its capacity to toggle between old and new power is well honed: its leaders know how to engage with old power tactics, and when to get out of the way and let new power surge.

Bloomberg's organization, for all its funding and public support for sensible regulations, was unable to rally a crowd around the outcome it desired. And as long as this was "Bloomberg's" campaign, it was hard for anyone else to own it. The NRA, on the other hand, shows a rare capacity to be dramatically present in one moment, and to disappear entirely the next. It's instructive that after the failure of the Manchin-Toomey bill Bloomberg pivoted, launching Everytown (with help from Jeremy's organization, Purpose), an effort to build a grassroots national movement for gun safety. The group includes a cadre of activist women who had formed chapters after the tragedy at Sandy Hook under the banner "Moms Demand Action for Gun Sense in America." These women carry the moral authority of being mothers who are passionate about keeping their kids and communities safe. Their protests and savvy messaging have pressured big retailers like Starbucks and Target to ban guns in their stores. And in recent years, a more grassroots and locally focused Bloomberg operation has achieved significant policy victories, such as winning ballot initiatives on background checks in Washington State and Nevada.

Everytown understands the need to close the "intensity gap" with the gun rights folks. A sign that its strategy may be working is that, after the mass shooting at the Pulse Nightclub in Orlando, Florida, Everytown supporters, which by then numbered three million, made 62,000 calls to the Senate in one week, nearly eight times

the number gun control supporters had made in the months leading up to the failed Manchin-Toomey vote. Future battles between
Bloomberg's group and the NRA may be more evenly matched.

The NRA draws on the strength, gravitas, and resources of a big
institution and the energy and creativity of a social movement. It
uses blended power in a big, messy, multi-player space. The next
story shows how TED blended power to build something more self-
contained.

OPEN AND CLOSED: HOW TED CHANGED
THE WAY IDEAS SPREAD

What TED has in common with Tiffany's

Every second of every day, seventeen people start watching a TED
Talk. But for more than fifteen years even the biggest TED Talks had
an audience of just a few hundred. TED was a small group of fancy
people gathering in Long Beach, California, once in a while to hear
enchanting lectures about technology, entertainment, and design
(hence "TED").

TED, which became an annual conference in 1990, was always
a Cheerleader and incubator of new power ideas. Early talks were
on topics like "the new power of collaboration" and "the birth of
the open-source learning revolution." But its model was anything
but open source. The program and speaker lineup was curated
by a small, centralized team. The people who attended TED had
privileged access to some of the most cutting-edge thinkers in the
world—but that access was restricted to those with enough means
and the right connections. As recalled in the official TED history,
the early framing of the conference reinforced the idea of it as a
kind of secret society: "a growing and influential audience from
many different disciplines . . . united by their curiosity and open-
mindedness—and also by their shared discovery of an exciting
secret."

In the 2000s, with the opening up of the digital world and the rise of social media, you might assume that TED would have abandoned this closed-shop elitist model, disrupting itself to embrace the collaborative, democratized, connected world that its speakers so often imagined.

Yet what is so interesting about TED is that it didn't change its core conference model. In fact, in some ways it has become more inaccessible than ever. Attendance at today's flagship TED event, which takes place in Vancouver, will set you back as much as $25,000. Even with that price tag, conference-goers are hand-selected, just as the speakers are. And there is hot competition for the 1,200 places and a swag bag full of the best shiny Silicon Valley bling. Only a happy few get to rub shoulders with Al Gore or play Cards Against Humanity with Sergey Brin. This is thought-leadership as a luxury good, hardly the stuff of mass participation.

The great irony—and achievement—of TED is that while all that is true, it has still become one of the most loved, shared, popular, and respected new power brands in the world. That part of the story begins with a British entrepreneur named Chris Anderson.

Anderson took over the TED conference in the early 2000s. As its new leader, he went out of his way to respect the unique culture of TED, shifting it to become a nonprofit while calming the fears of dedicated TEDsters that he might screw up their well-loved vibe. Yet the talks he heard at each convening underscored how much the world was changing. And he knew TED must change, too. But rather than tinkering with the core model, he looked outward, seeking to build communities around the conference. If he succeeded, each would create value in its own right; each would feed value back to the core.

He began with a big, and seemingly risky, decision. TED would post its talks—essentially its sole intellectual asset—online and for free. Former top TED exec June Cohen has said that there was real resistance to this from some at TED who feared that posting the talks online would undermine the business model of ticket sales.

But they saw precisely the opposite effect. The first talks were posted in June 2006 and by September they had racked up more than one million views. Today that number is in the billions. The hugely increased awareness and buzz around the brand increased demand for tickets and drew in an array of corporate sponsors who lined up to support TED's work. The more open TED was in sharing its talks, the more valuable its closed convening became.

TED, of course, did a lot more than simply post some videos online. What made the talks so popular is how carefully designed they are for the mass market. Beautifully shot, effortlessly sharable, and strictly timed to last no longer than eighteen minutes, they are tailor-made, in Anderson's phrase, "to illuminate, clarify, engage and delight." (A familiar critique is that TED "dumbs down" important content, though increasingly an eighteen-minute format feels like a pretty generous allocation of attention.) The promise of the TED talk is that you become more interesting by watching them, and you appear smarter to your friends by sharing them.

What works so well for TED is that these two user experiences—the ultra-VIP attending the Vancouver conference and the ordinary person sharing one of the talks—rely on a similar set of incentives, just on very different levels.

Like the jeweler Tiffany, which can whisk a billionaire upstairs to a private salon to inspect the rarest diamonds while it welcomes selfie-snapping tourists in its lobby showroom, TED is able to offer a select few access to the very top of the market and a little something for everyone else. This bifurcation has largely reinforced the value proposition for both sides. Those who speak at the conference have a much more valuable experience because they know that the talks have an influential mass market. Equally, those viewing the talks on their smartphones know that they are accessing a high-end quality product because they see TED as the most credible curator.

Had TED remained a purely closed shop, it is hard to imagine it would have retained its relevance. At a time when the world is looking skeptically at the hidden dealings of global elites, its annual gathering would have appeared increasingly anachronistic. (Note that the World Economic Forum, whose cloistered convening takes place annually in the ski resort of Davos, Switzerland, is facing a similar challenge, and has responded by building communities and events beyond its annual meeting.) But Chris Anderson's digital play delivered TED a global audience and a nascent community. In effect, TED offset its old power with new.

How to scale exclusivity

By opening up its talks, TED created the space for the sharing of its content. But as it became a household name, its success led to an unexpected challenge. The many new TED enthusiasts were hungry to do more than just share and add comments. They wanted to deliver their own TED Talks. Obviously TED's annual gathering could not meet this need, so Anderson and his staff invented TED's next community, which they called TEDx (with the *x* standing for self-organized).

TEDx works by allowing organizers everywhere to channel their inner Chris Anderson. Today there is TEDxAmsterdam, TEDxThe-BrowningSchool, even TEDxInyaLake in Myanmar. Thousands of events have taken place, with over 30,000 TEDx talks delivered in 130 countries around the world.

Lara Stein, the woman charged by TED with founding and running TEDx, describes the moment that big TED, which had to some degree remained wary and culturally resistant to TEDx, began to see what it had on its hands. In 2012, six hundred TEDx organizers met for a week in Qatar for the first TEDx Summit, at which they shared experiences and dreamed together about what the platform could become. "It didn't matter if you were a kid out of a shanty town in Kibera or a Carnegie, you got to come and have this experience," Stein recalls. "The bonds that were forged there carried the movement for many years."

But increased access to the brand has not meant abandoning control. Like the conference in Vancouver, the TEDx experience is carefully structured. In the official "rules" for wannabe TEDx organizers the word "must" appears forty-eight times. There are twenty-seven appearances of "should" and twenty-one of "cannot."

The structure of TEDx is less similar to a distributed movement like the hacker collective Anonymous—where anyone can claim membership and take things in whatever direction they please—and more like a franchise, where a service agreement enforces specific expectations. The parameters are designed to preserve what TED sees as intrinsic to its model: high-quality curation, the absence of a profit motive, and a ban on pay-to-play sponsors. TEDx organizers must make a clear case for why they are well positioned to pull off a compelling local event, and they need to play by TED's rules.

Occasionally those rules are broken. Nilofer Merchant, writing in the *Harvard Business Review,* has documented various skirmishes over TEDx talks that peddled pseudoscience, or technical-sounding nonsense topics like "vortex-based mathematics," "plasmatics," and Egyptian psychoaromatherapy. The last two of these talks were at TEDxValenciaWomen in 2012, an event described by one attendee in Merchant's account as "a mockery . . . that hurt, in this order, TED, Valencia, women, science, and common sense." The controversy around these talks led to a moment of "shaping" by big TED's leadership: it moved to realign the TEDx community with TED's mission, providing detailed information and new guidelines on how to carry this out.

If the first reason TED is so prescriptive is an effort to protect its brand, the second is an effort to extend it. Key to its success is the paradoxical idea that they can scale exclusivity. On a smaller level, they want every TEDx to mimic the dynamics that make big TED such a hit. Most TEDx organizers are prepared to accept these restrictions. "The reason behind that is most people who get started have no idea what it's like to coordinate a conference. It's complex. They want to protect the brand so somebody doesn't throw a lousy party and then everybody says, 'Well, I went to a bad TEDx event,'" said Randy Bretz, the passionate seventy-something organizer of TEDxLincoln. He went on to say that "the TED brand is phenomenal. In my mind, that's the key. If it was Randy Bretz doing an idea-sharing conference, I don't think it would go very far, even though I have a lot of friends and connections . . . Now my business card says TEDx Curator on it and it's a source of pride for me . . . My

wife would put it a little more bluntly: 'Randy thinks he's important because he's doing this.' She also says, 'I'm extremely proud that
you've gotten involved in this. I think it's a great thing.' "

TED's Open Translation Project formally launched the same
year as TEDx. But this fourth TED community began organically,
with people around the world either asking if they could translate
their favorite TED Talks or simply translating them without asking.

The platform TED launched in response to this interest was
simple and elegant. Translators would post their translations, other
translators would act as "reviewers," and a "language coordinator"
(whose role was to oversee all content in a specific language: French,
Spanish, Urdu, and so on) would be the last set of eyes on it before
it was published.

The project has now seen almost 100,000 translations published
in over 100 languages by over 20,000 volunteers. Some translators
engage one another online; others even get together in person. All
enjoy the opportunity to broaden access to talks that they think
matter. These will often be the talks from an event they organized;
other times a "hidden gem" they are excited to share more widely.
The most prolific translator is Sebastian Betti, who also curates the
TEDxPESA event in Argentina. He has racked up over 3,000 translations, reviews, and transcriptions of talks.

Open vs. closed systems

Taken together, TED's four communities offer a master class in how
to blend power under one unifying organizational brand. The three
new power worlds it has built around its old power conference have
turned it into one of the most effective media platforms in the world.

If we contrast the blended power of TED with the blended
power of the NRA, we see one big difference: TED is essentially
a closed system. Every interaction is circumscribed by the brand.
Though its putative mission is "ideas worth spreading," its leaders
have not made the decision to support just *any* good idea worth
spreading, no matter what format it is in or where it was delivered.

Theirs is a huge—but closed—ecosystem of ideas, with the bright red gleaming TED brand tied to every single piece of content and event produced.

The NRA operates inside an open system. Yes, it has an old power brand it protects, but it also fosters a wide ecosystem outside of its core organization and membership, and has a much looser and less circumscribed relationship with that broader community.

These are two very different approaches, each of which serves its organization's core goals. Through its open system, the NRA can draw on the energy and intensity of gun culture and commerce. And through its closed system, TED protects quality and strengthens its product and brand value.

Up until now, this singular focus on brand has been very good for TED. The term "TEDsters," once reserved for the select few who attended the annual TED conference, is a marker of identity around the world. But TED may yet have to open up its world even more. The conference itself is vulnerable to being tarred with the brush of increasingly unpopular Silicon Valley elites. And the TED format— strictly focused on producing unexpected and delightful talks—can feel inadequate to TEDx organizers who want to turn their events into opportunities for real-world action and year-round community engagement. TED hasn't always been comfortable about taking this next step. Ideas worth spreading is one mission. Problems worth solving is another. At a time when the crowd's desire for participation is only growing, TED may have to reimagine itself once more.

HOW CROWDS CAN BUILD CARS: THE EXPERTISE OF EVERYONE

If there is an industry most identified with old power and twentieth-century management, it has got to be the auto industry. And yet our next story is of a company that is making cars in a distinctly twenty-first-century way, one that relies closely on blending old and new power.

"We do our best thinking and creating behind open doors." That slogan is one of the rallying cries of Local Motors, and it announces a very different kind of car company.

The *Financial Post* describes Local Motors' business model as "so radical that it's hard to comprehend at first: crowd-sourced, 3D-printed electric vehicles built in local microfactories the size of grocery stores, then sold directly to consumers." Local Motors has a community of 50,000 people not on its payroll who design and develop concepts for vehicles that the company makes and sells in small batches. Think artisanal jam, only with engines. And it sometimes assembles these vehicles with the help of those community members. This is the opposite of Fordism, the last century's reigning mode of manufacturing, in which standardized products are made in great numbers on assembly lines and then sold to a mass market. Local Motors offers its customers the antithesis to Henry Ford's famous offer: "Any customer can have a car painted any color that he wants so long as it is black."

The Local Motors community is made up of enthusiasts, from the highly skilled to the amateur, all of whom want to dream big about the future of cars—and play a part in making that future happen faster. Together, members of the community have designed and developed a self-driving minibus, a special vehicle dedicated to the art of transporting pizza (with built-in oven!), a souped-up, road-legal monster truck, and more.

To accomplish this, Local Motors and its young co-founder Jay Rogers have flipped a top-down industry on its head, drawing on its customers for direction: "We asked you, the Local Motors Community, what you want to improve in your cars, and you said you want to be able to do your own automobile maintenance, repairs and modifications. You've said that today's automakers make it difficult to access parts and perform simple tasks like replacing headlights and charging dead batteries. You've said that identifying rattles and noises is difficult because parts are buried within the inaccessible depths of your cars' structures. It's like we're all driving mystery machines."

Here's how Local Motors framed one of its most recent challenges to its community: to imagine the world's first highway-ready 3D-printed car, with open-source design and parts so that anyone can repair it, and a modular concept that allows the owner to create a vehicle to his or her own specifications.

The winning design was the "Swim," the brainchild of Kevin Lo of Vancouver (username Reload), a Hewlett-Packard engineer who had taken to learning about car design in his spare time to get better at the user-interaction aspects of his full-time job. "I actually kind of styled it after a woman's shoe," he said.

Local Motors' design philosophy is the polar opposite to that of Apple, where industrial designers are considered all-knowing, even God-like, and where the company assumes that the consumer doesn't know what it wants until the product descends from the heavens of Cupertino.

But what is driving Local Motors isn't anarchy. Its success is in the careful blend between old and new power that allows its crowd a substantial (and communal) role in designing its cars, then makes sure they roll off the line.

Alex Fiechter, Local Motors' director of product development, points out that the company has staff from all the same technical disciplines as its more traditional competitors, and a similar product life cycle. Yet it has found a way to open up the process to a much bigger community of people.

After the design for a vehicle like the Swim has been chosen, Local Motors posts a set of other "project needs" to the community, the same basic tasks any car company faces as it tries to turn a design into reality. Fiechter explains, "We're looking for those serendipitous interactions where we virtually bump into somebody in the hallway talking about the problem that we're running into right now, and they say, 'Yeah, actually in my previous job, or maybe in my current job, I'm working on this.' They are looking to find that member who will respond something like: 'This is my hobby. I'm all over this topic. Let me tell you what you need to do.'"

The best of both worlds

Jay Rogers says his instinct is always to turn to the community. But his success is in knowing when the company needs to set the terms and make the right call. He calls his approach "drawing the box tightly," setting clear parameters and conditions within which the community can create and ideate. The tightness is guided by the company's view of what is possible in the supply chain and its knowledge of the full range of external conditions shaping the viability of a particular product.

When Kevin Lo submitted his entry to the challenge, Local Motors' design community liaison at the time, Nicolas de Peyer, initially rejected it for not meeting the specifications of the challenge. "I wanted to validate this so badly and it is so close—the only thing that is holding me back is that it's not very clear about the soft top convertible part of this," he told Lo, adding details of what Local Motors needed to see. When Lo jumped through these hoops, de Peyer accepted his entry and added a hearty congrats. This kind of detailed back-and-forth is typical of how Local Motors structures its community's participation.

Rogers acknowledges that the community challenges him back: "There wasn't a single critique that we hadn't heard, 'You don't have thirty-five-inch tires, you have thirty-four-inch tires, you should have had thirty-five-inch tires.' Yep. We heard it, we thought about it, we respected it, we realized that there's a trade-off. Thirty-five-inch tires are $2,000 higher, thirty-four-inch tires are one inch smaller and so you lose some material performance . . . but you can buy each one for 150 bucks. We made a decision not to put $2,000 tires on the vehicle." When you explain that to the community, Rogers said, "they'll say 'Well, it does make sense, I never knew that.'"

The community has an expectation of a good dose of old power from the company, too, once in a while. "Sometimes we've got people that have just said, 'You're asking us to do too much.'" For example, when it turned to the crowd to help design a web portal, there was a protest from some members: "You guys do any work?"

Local Motors is now figuring out the best way to share the value of its creations with its community. Community members already own their contributions to the platform through a Creative Commons license, one that agrees to share that intellectual property if the entry is chosen. Winning submissions earn prize money, and in at least one case, the winning designer is earning royalties from his creation. Rogers sees giving royalties only to the most important contributors as the best long-term path to sharing value; he feels that a more transactional model in which every contribution is paid for would tarnish the spirit of cooperation and volunteerism that binds the community together. All this engagement pays off when it comes time for Local Motors to bring its cars to market. Here we see the "participation premium" in action. People who have been involved in the development of the car are primed to both evangelize for and buy them. The community Local Motors has built is a free marketing channel to rival those of cult brands like Harley-Davidson or Tesla.

It's still early days for Local Motors. The company is far from having the scale of the big automotive players. But it shows us what a true crowd-powered manufacturing business might look like.

Our final story takes us to Spain, for the tale of a group of citizens who didn't wait for old power politicians to tap their rising energy. They blended power to break their way into the heart of political life.

THE WILD RIDE OF PODEMOS: UNPACKING THE PRAGMATIC REVOLUTION

Stop us if you've heard this one before.

A young charismatic leader emerges unexpectedly to find national political prominence. His mastery of digital tools allows him to outflank the opposition. His dazzling oratorical skills inspire massive rallies. Relatively inexperienced, he galvanizes a generation with the battle cry "Yes, we can!"

This is the story not just of Barack Obama, but also of Pablo Iglesias, the leader of Spain's Podemos party.

Podemos, which translates to "We can," was founded in early 2014. Just four months later, in the European parliamentary elections held in May, the party received 1.2 million votes, 8 percent of the total. In the Spanish general election in December 2015, a full fifth of the vote went to its candidates. In less than two years, Podemos became the second largest Spanish party by membership, breaking up decades of two-party rule.

The party's unpredictable and unpredicted rise tells us a lot about how to blend power. It's a story of an organization committed to new power principles and values, but not at any cost. What is so striking is how often, and how fully, it deploys old power strategies to get the job done. Its philosophy—as a party leader, Jorge Moruno, has commented—is part Don Quixote, part Machiavelli.

The Podemos story begins on the back of a big disappointment (and a familiar lesson in the limits of new power). The spring of 2011 was a tense time in Spain. Corruption was rife. Bankers were being cast as villains in the collapse of the economy. Youth unemployment had reached a staggering 43 percent. Public squares were full of demonstrations. Out of the tension emerged the Indignados (or 15-M) movement—a mass protest of Spain's budget cuts and a precursor to the broader Occupy movement that would rattle the world that fall. It was new, exciting, and full of possibility. As *The Guardian* reported at the time, "City square by city square, individual meeting by individual meeting, thousands of citizens have come together in a networked approach to politics that is fresh and engaging because it defies, above anything else, the hierarchical approach favored by vested interests."

Yet all this political energy didn't change much at first. In the November national elections, the two main parties came out on top, just as they had since the 1980s. Looking back, Pablo Iglesias

remembers the "sneering response of the old elites," who would lec-
ture his fellow demonstrators in the squares that if they had better
ideas, they should run for office.

Iglesias, long-haired, lithe, and loquacious, then in his mid-
thirties and a lecturer in political science at the Complutense University of Madrid, heard these sneers as a challenge. A lifelong
activist committed to an egalitarian Spanish society, he had been
named for the father of Spanish socialism and had been a member
of the Communist Youth Union.

So, along with a ragtag group of fellow professors and students,
he began to make plans to build a new leftist movement that would
not fizzle out like 15-M had. Instead it would grow in intensity and,
when it was ready, translate its energy into that of a formal party. It
would then win enough political power to change the electoral cal-
culus in Spain. And he began this fight with a very old power move.

He started a TV show.

An old power beginning

The show was called *La Tuerka* (The Screw) and was delivered via
YouTube. Low quality and poorly produced, with cameramen scur-
rying in and out of the shot, and guests sitting oddly close together,
it was experimental to say the least. But it was the perfect boot camp
for something Iglesias and his academic and activist allies believed
would be a necessary condition for political success: the capacity to
share their messages via traditional media.

Sure, Twitter would get the younger folks buzzing, and Facebook was great for mobilization. But to make big cultural shifts that
might translate into lasting political power, you needed to occupy
(their word for it) the airwaves. For Iglesias, "TV studios have
become the real parliaments." Their revolution *would* be televised.

After three years of practice on *La Tuerka*, Iglesias began to step
out onto the mainstream shows. His now-polished message was a
welcome jolt for the major media. Soon he was being booked on

every network; the ponytailed professor as symbol and salesman for the plight of ordinary people. As he described it, "Step by step, an unconventional left-wing talk-show guest became a reference-point for the socio-political discontent caused by the crisis."

At this point, despite their new power rhetoric, the focus was all on a figurehead, Iglesias. Now, when you are dynamic and passionate and look almost exactly like a 1950s Hollywood Jesus, it is probably quite hard to resist becoming an icon, but this was an intentional and strategic choice. As Iglesias described it, this kind of fame was "the indispensable ingredient of the Podemos hypothesis" and its quest to win a place in the homes, hearts, and routines of everyday people.

The next move was to create an identifiable group of people to blame. So they began to turn the people of Spain against *la casta* (the caste). "La casta" was a new blanket expression they designed to label those seen as responsible for, and benefiting from, the country's ongoing crises. In one memorable term it nailed the bankers, the corrupt businessmen, the biased judges, and the dodgy administrators. It captured the Punch-and-Judy show that was Spanish politics, where, no matter who won, the same elites always ended up on top. Soon, across the country, *la casta* signified not just a crooked class but a crooked system.

By the end of 2013, Iglesias was everywhere. So were his slogans and arguments.

They had a hero. They had a villain. It was time to form a party.

A new power campaign

Old power tactics and a charismatic leader had built the energy that made it possible, but the formal launch of Podemos came with a flurry of new power.

In early 2014 a core group of academics and activists came together to formalize their thinking, set their goals, and churn out a political manifesto, Mover Ficha (Making a Move)—something of a love letter to new power values. It called for "horizontality and transparency," "democratic openness," "sovereignty of the people," and a candidacy that results from an "open participative process."

Iglesias announced their work would only continue if 50,000 Spaniards signed a petition of support for their campaign. If that number was reached, he would run for the European Parliament and the party would create an open process for others to become candidates under the Podemos banner. The number was reached in less than twenty-four hours.

And although Podemos was now formally a party, it kept on behaving like a movement. There was still no particular effort to institutionalize or add a lot of structure. Podemos's electoral strategy was largely to do all it could to encourage people to grab the campaign and make it their own. It put out a call for people to organize and create "circles"—locally and online—to discuss the concerns they had and share ideas. Circles began to spark up all over the country.

Eduardo Maura, part of the Podemos team working toward the elections and Iglesias's fellow lecturer at Complutense University, called this the "hacker logic" of the movement: "When you are doing politics as a hacker you proliferate, you have to be everywhere, you want to be everywhere." This thinking drove their low-barrier, light-lift affiliation model, the polar opposite of most twentieth century political parties. They made connecting with the party as easy as possible. "To create a branch you only need a Facebook account, an email and a meeting. No membership, no fees. So, in the first two months we got more than 300 branches, not only in different places but also branches that had to do with specific fields, like education, culture, the environment." There was a heavy emphasis on local agency in the way the circles were constructed: people were creating Podemos circles not based on what the "head office" thought was important, but around the issues they cared about most.

Podemos respected and understood the diffuse, fleeting, and untamable nature of a digital movement. The central figure of Iglesias and the central theme of injustice would guide their direction, but the energy was now broadly distributed.

This strategy paid off. Five candidates, including Iglesias, became members of the European Parliament. Yet perhaps the bigger meaning of the win was as a demonstration of the party's political viability. The real-world victory validated all those citizens who had added their vote, their voice, or just their likes to the movement.

A blended-power party

By any measure, Podemos had arrived. But party leaders wanted to go much further, building on the mass participation they were known for, by channeling citizens' energy into a structured approach to policy- and decision-making.

Their big idea was a two-month-long Citizens' Assembly that would allow "popular participation in the most important decisions." Through online and off-line discussions, public meetings, and meetings of the circles, Podemos's goal was to crowd-source key aspects of its manifesto and policies and choose leaders. The process was a little cumbersome and uneven, but certainly very open. Anyone was welcome to submit drafts of their ideas. *The Nation* called it an "unprecedented exercise of direct democracy" with over 100,000 people taking part. An online debating site, Plaza Podemos, launched, attracting 10,000 to 20,000 followers every day. A "participation platform" allowed people to fund the party's efforts, either through contributing to its general work, making microloans to back election campaigns, or supporting specific projects.

For all of the new power involved in the process, it delivered something of an old power result. Iglesias's central role was solidified by his election as general secretary, and his platform of proposals won the day. What's more, he successfully fought off a proposal that the party have three general secretaries, as a check against concentrated power at the top. More than ever, even after an extensive participatory process, he was the dominant face and leader of the movement.

By December 2015, Podemos had built such strength and momentum that it ended up winning over five million votes and sixty-nine seats in Spain's parliament. It had broken the country's two-party rule before its second birthday.

Four months before the Indignados movement broke out, two thousand miles away Wael Ghonim was launching his own assault on the establishment. He was an administrator of the "We Are All Khaled Said" Facebook page that sparked the Egyptian chapter in the Arab Spring.

Yet this surge of new power, like the Indignados and Occupy movements, would dissipate. In Egypt, the unrest of the revolution presented opportunities for old power forces—more despotic than those they replaced—to grab control. As Ghonim recalled in a 2015 TED Talk, "I once said, 'If you want to liberate a society, all you need is the Internet.' I was wrong."

Ghonim, looking back, believes that the distributed leadership and social networks of the Egyptian protest movement actually served to weaken rather than strengthen it. "The euphoria faded, we failed to build consensus, and the political struggle led to intense polarization. Social media only amplified that state, by facilitating the spread of misinformation, rumors, echo chambers and hate speech."

What Podemos teaches us is how to avoid the fate of fizzle. Of all the organizations in this chapter, it began with the greatest ideological commitment to new power. Yet its success depended on constant pragmatism. They were careful not to be overly dazzled by the promise of social media, nor live on the adrenaline of protest alone. Founders were always willing to pivot to old power when the situation demanded.

Podemos is not alone in successfully blending power to mount a political insurgency. In neighboring France in 2017, Emmanuel Macron swept into office by launching a movement, En Marche, that very quickly transformed itself into a governing political party. He did this by skillfully exploiting dissatisfaction with establishment politics, rallying digital crowds, and marshaling mainstream media attention to build an organized political base in record time, something that would have been much harder to do in another era. Yet he did all this running as a fairly conventional centrist who has not embraced the Podemos brand of radical participatory politics.

Of course, there are challenges inherent in the Podemos approach. We should always think twice when charismatic leaders claim to represent the voice of the people, and Podemos remains overly reliant on the force of Iglesias's personality. There is risk, too—which Podemos is well aware of—that idealism in campaign mode will turn to entitlement and cynicism when it governs. Placards can be perfect; policies never are. The party has seen successes, especially in its influence on cities, with mayors in Barcelona and Madrid emerging with the support of the Podemos movement. Yet its elected officials' early years in parliament have been made up of concessions, coalitions, squabbles, and compromises, all set in the maelstrom of a new multi-party reality.

But the party's initial achievement is undisputed. Spanish politics has changed, becoming younger, more plural, and more participatory as a result of its insurgency. And Podemos has created a model for a new kind of political party that the world needs.

Those who master the skills we have laid out in this book will be able to shift between old and new power as situation and strategy demands: to flip from open to closed, to toggle between "movement" and "institution," to know when to control and when to release control.

There is urgency around this work. If we want to imagine a more positive and participatory politics, Podemos is instructive. The most important causes in the world—though they might bristle at the comparison—would advance their work further with the blending capacity of the NRA.

Many of us understand how to speak old power—after all, that is the world we grew up in. Rising generations may have new power as their mother tongue. But those who really change the world will become fluent in both.

NEW POWER AT WORK

Reuven Gorsht, global VP of customer strategy at SAP, told us a funny story about an intern he hired a couple of years ago. She did the job for three weeks, then mentioned in passing, "Oh, you know, I met Bill."

Bill is the CEO of SAP. Reuven had worked there for ten years and had "never had a minute with him."

"Oh, that's cool," said Reuven, studiedly nonchalant.

"Well, I've got a meeting with him next week for a half hour."

Reuven was a little flummoxed. What had just happened?

His intern explained: "I came up to him and I said, 'I'd like to meet with you to give you feedback.' And he said, 'Fine,' and gave me a time."

Many people have their own version of Reuven's story. Of all the challenges in the workplace, perhaps the most mystifying for the old power leader is the daily task of management in a world where many now have new power values. Younger workers, and others as well, are increasingly pushing back against norms like hierarchy, loyalty, and the very idea of the specialized professional who stays in her lane. Old power and new power are colliding in today's workplace.

In the twentieth century, the world of work was arranged according to the color of your collar. Blue-collar workers formed unions and fought and won basic rights and entitlements designed to protect them from the often insecure nature of their trades. White-collar workers were soldiers in the great march toward managerialism, a governing ideology seen everywhere from the League of Nations to Ford Motor Company to the Japanese "company man." The role of "the professional" was to slot precisely into elaborately managed and ever-larger organizations. In return for what would often be decades of steady service, he would steadily move up the ranks and be rewarded with job security and social status.

Many of our old power leaders grew out of a world that prized those who fit well into a set process and organizational structure. Yet today, that exact skill can be seen as a major liability.

In the twenty-first century, managerialism as both ideology and practical reality is under assault. Our cultural heroes are increasingly scrappy disrupters, not corporate managers; the efficiencies of managerialism are seen as sand in the wheels of innovation; decades-loyal "company men" are being replaced by contingent and on-demand workers; and an amorphous "maker culture" is challenging long-standing norms about expertise. Old power managers are facing workplaces that can feel like they are full of wannabe Elon Musks, with vastly higher expectations and unending demands for feedback, with one eye on their next promotion and the other on their next job.

Too often, this tension gets lampooned as Old Codger versus Young Turk. AARP versus ADD. But there is a deeper cultural shift playing out as old and new power values do battle at work. In this chapter, we'll unpack why new power workers are approaching the world so differently, and we'll suggest how managers can deliver the recognition, opportunities, and agency required to get the best from them (without losing their minds in the process).

NOT WHINING BUT PLEADING: WHY FEEDBACK
AND RECOGNITION NOW MATTER SO MUCH

Just imagine for a moment that your workplace functioned with the same dynamics and incentives as a social network like Instagram.

In that world, an ordinary act—let's say submitting your expense report to your boss—would be met with some pretty extraordinary validation.

Within seconds of hitting Send on your report, you would receive dozens of comments from your colleagues saying how terrific your work was. You'd hear things like "OMG that is AMAZING!!!!" "So cool!" "Great work!" People would stop by your desk to smile at you and congratulate you on getting it in on time. Someone might even bring you a little heart-shaped sticker to wear on your shirt. Your boss would share copies of your expenses with everyone he knows with a note that says "#MUSTREAD!"

Throughout the day, you would enjoy a steady stream of praise, as more and more people became enthralled by how much you'd spent on taxis and cappuccinos. You'd get notifications that say things like "Pradeep just read your amazing report!" Your expenses wouldn't actually need to add up. After all, the only people assessing them would be those predisposed and hand-selected to affirm you, offering dose after dose of personalized delight.

Contrast this experience with something that looks a lot more like reality.

A twenty-three-year-old assistant spends all weekend on a new marketing idea he has for his company. He is super-excited about it, and goes above and beyond to put together a slide-deck that lays out his thinking. He emails it to his boss first thing Monday.

And then he waits.

On Wednesday he sends a follow-up, just to check she's received it.

On Friday, he gets a note back apologizing, saying she's been at a conference, and will get to it next week.

The Thursday after that, she finally emails back, saying, "Interesting . . . Thanks for sharing."

It is both easy and common to cast a new generation of workers as narcissists: whiny millennials desperate to spend every moment talking about themselves and their feelings, and expecting a medal for doing so. But it is smarter to see these expectations as signals that the workplace has failed to create the satisfying feedback loops that have multiplied in other aspects of our lives. Research from the Young Entrepreneur Council shows that 80 percent of millennials would prefer to get feedback in "real time." A 2014 Millennial Impact Report showed that "more than half (53%) of respondents said having their passions and talents recognized and addressed is their top reason for remaining at their current company."

There is a reason that feedback and recognition are so highly prized by the new power set. For most, their lives are punctuated (perhaps even defined) by the validation and engagement of others. Every text, every image, every post is a call designed for a response: the drip-drip-drip of dopamine-reward they receive from every heart-shaped like. Even their dating lives are adjudicated by strangers looking at carefully curated photos and deciding to swipe left or swipe right. Being judged, in explicit and implicit ways, is part of their lives. It is how they navigate, guided by a galaxy of digital North Stars that shape, encourage, and catalyze every move.

Their lives are rich in data, too. They measure their steps, calories, heart rate, and sleep. Think of the experience the millennial worker has as he runs home from work using the Nike running app. He gets instant information on his speed and pace, dressed up with a nice map of the route ahead. He can "get cheers" from his Facebook community who can track his progress in real time and applaud. He can compare his times with his previous times and those of his friends. Once he completes his run, his stats are posted to his community for their validation and admiration. Then an Olympic gold medalist praises him via his Bluetooth headphones to tell him what a great job he's done. (Believe it or not, there is now even "wearable" technology that monitors a person's energy levels in the bedroom. On completion, a mobile app offers you a score.)

Indeed, in our digital lives, we are increasingly conditioned to get badges and status upgrades for doing little more than showing up. One of Snapchat's core features is its "trophy case," where you can display the trinkets you have been awarded for such Olympian feats as "Verified your email address in Settings" and "Sent a snap with one filter applied."

Of course, for many young workers this is the story not just of their digital lives, but also of their childhoods. They were raised as part of a "trophy generation," recognized for the act of participation rather than the level of performance, getting medals and lavish praise just for completing the race. The parenting philosophy that governed their early years—build self-esteem above all—conspires to deliver the same artificial feedback loops as most social networks: persuading the individual that the ordinary moments of their lives are actually of extraordinary value. Those who enter the workplace with this mindset often find that reality bites hard. And so do their bosses.

Emerging approaches are helping organizations meet this new power worker on her own terms. The online service TINYPulse is a user-friendly solution that asks employees one simple question each week. This provides two benefits. First, it offers an outlet for those expecting a chance to share their views. Second, it provides managers with regular doses of data that can be aggregated, analyzed, and, ideally, acted upon. Playful and friendly, the "voice" of TINYPulse sounds nothing like corporate-speak. Questions like "If your company were an animal, what would it be and why?" line up next to "Do you intend to be working here in 12 months?" It is a far cry from the traditional staff survey that is rolled out annually and then disappears quietly into the graveyard that is the company intranet. And it only takes two minutes. It has been adopted by IBM, Facebook, and Airbnb, among others. When Henry experimented with TINYPulse at his organization, its "cheers for peers" feature really took off. A peer-to-peer digital recognition system, it lets everyone from the boss to the intern publicly recognize their colleagues for good work.

Products like TINYPulse offer a hint of what is to come: managers receiving and delivering feedback to new power workers in a way they're familiar with—drip-drip-drip style. But this is not to say that the burden is entirely on management; those with a new power mindset need to recognize the big difference between their boss, who is there to help them grow and develop, and their followers, who are simply there to cheer them on.

There is a big opportunity here. Having a workforce of people who *want* more feedback and agency is not something to be squandered. Note the opposite challenge in the old power workplace: employees—especially those who have been undermanaged for years—who can view any kind feedback as a criticism and an existential threat.

THE "FOUNDER FEELING": WHY (ALMOST) EVERYONE WANTS MORE AGENCY AND WHAT TO DO ABOUT IT

In 2015, MTV conducted a survey asking 1,000 young people born after the year 2000—the post-millennials—what they thought their new generation should be called. The winner? "The Founders Generation." It's an answer that tells us a lot.

In a new power world, where people are increasingly skeptical of traditional institutions, the great professional myth of our time is of the disruptive founder who builds a vast unlikely empire from scratch. The story of Evan Spiegel, co-founder of Snapchat, one of the most popular social networks for this emerging generation, offers a textbook example.

In the first chapter of this familiar start-up tale, Evan and two co-founders, Bobby Murphy and Reggie Brown, have their big idea. They will create a photo-sharing app that makes it fun to annotate and draw on photos. But with a catch: the photos will then disappear. (This was an idea well timed for a generation wrestling with the competing urges to protect their privacy and to send naughty pictures.)

Chapter 2 brings the obligatory dropping-out from a prestigious university. Spiegel checks this box by abandoning Stanford University only a few credits short of graduation to pursue his Snapchat dream. He then shifts neatly into chapter 3—a legal brouhaha with his co-founders—falling out with Brown, the guy who first had the idea. Brown sues him.

Our story hits its climax in chapter 4, when Spiegel gets approached by his entrepreneur-hero, Mark Zuckerberg, who offers to buy Snapchat for *$3 billion*. But will Spiegel believe enough in his idea to reject his hero and risk it all?

Hell yes.

You know what happened next. Snapchat grows wildly, Spiegel thrives. The company goes public at a $24 billion valuation. Spiegel marries an Australian supermodel.

Perhaps best of all Spiegel gets to tell much older people what to do every day, the dream of any young person. As one former employee told *Fast Company*: "He just wants [people] to listen and do what he's asking, and he's really getting sick of having to explain [what Snapchat is] to one old fart after another in the media industry."

Is it any surprise the founder myth has such appeal?

Of course, there's plenty of tinsel around Spiegel's founder narrative. Like Zuckerberg, he was born lucky, as he has noted himself. He grew up in the wealthy Pacific Palisades, the son of successful Harvard- and Yale-educated lawyer parents, with all the confidence that being rich, highly educated, straight, white, and male confers in America. (As has been well documented, that is precisely the founder profile that Silicon Valley tends to back again and again, in no small part because most venture capitalists are cut from the same cloth.)

Nonetheless, a generation of elites who would have previously coveted a partner-track job at Goldman Sachs or McKinsey are now fantasizing about launching their own machine-learning start-up called something like Splotchy. A Bentley University survey showed that 66 percent of millennials report they want to start their own businesses (of course, only a small fraction of that number actually do, and the rest are the people whose founder-like expectations are upending the traditional workplace).

More broadly, even if we don't dream of becoming a tech billionaire, we are now operating in a world where, across most facets of our lives, we're having founder-like experiences. As we are more and more the creators of our own myths, leaders of our own communities, stars of our own shows. New power dynamics make the desire for the "founder feeling" much more widespread.

What does it mean to want to feel and act like founders, even if we work for someone else? To answer that, it is helpful to think through what founders actually *do*.

- **Founders build things.** Founders create something new. Simply overseeing or improving someone else's initiative is no prize. Their validation comes from what they create.
- **Founders feel real ownership.** For entrepreneurs, this means owning equity (and then being sued by your spurned co-founder, if things go *really* well). In the new power workplace, it means a meaningful share in the value created by the organization. It also means getting to make your own decisions—when you're a founder, no middle managers are there to scold or micromanage you.
- **Founders experience the joys (and terrors) of transparency.** They know how much everyone is being paid, how much money is in the bank, whether the business paid its taxes on time. These are precisely the same kinds of demands we see from younger workers, who increasingly feel entitled to the kind of information usually reserved for senior executives or even CEOs, from full salary transparency on down.
- **Founders do many jobs.** When you're a start-up founder, you don't have the luxury of doing just one highly specialized job. In the first year or two of your venture, you're probably doing everything from buying cleaning

supplies to sales to putting together a new logo or
website. This mirrors the increasing demands in today's
workplaces for people to play roles that go far beyond
their more narrowly defined job descriptions—and to
play across disciplines and skills. There's an excellent
chance your new marketing hire is learning to code on
her lunch break. It is the era of the "side hustle"—
a recent survey showed that almost a third of millennials
now have an extra job on the side.

- **Founders are all-in and always on.** They benefit from
  greater flexibility and autonomy than nine-to-fivers, but
  the boundaries between their work and their lives tend
  to be blurry.

Of course, there is a dark side to this founder feeling. The one
part of the founder's experience that most people do *not* want to
share—yet do—is the uncertainty and insecurity. The reality that
three out of four venture-capital-backed start-ups fail mirrors a
workplace where people no longer feel secure in their roles and
aren't seeing the kind of long-term commitment from employers
that loyal employees could count on in the postwar era. Many are
now searching for firmer ground. A survey by ManpowerGroup
showed that job security was second only to money as the millennials' greatest motivation in choosing a role.

All these desires and anxieties are placing great strains on the
old power model, which delivers very little agency, information,
engagement, and flexibility for the worker, who largely has to fall
into line.

But some organizations *are* successfully building this "founder
feeling" into the fabric of their workers' daily experience. The pioneering Dutch network of home-care nurses known as Buurtzorg,
for one, has given a new power makeover to a traditional profession.

THE FOUNDING SISTERS: HOW SMALL AND MIGHTY TEAMS OF NURSES ARE REIMAGINING COMMUNITY HEALTH CARE

If you wanted to learn something about the future of the workplace, you probably wouldn't begin your quest by heading to the Netherlands and inquiring about the state of nursing.

But Buurtzorg—a decentralized network of nurses who engage with their communities on their own terms—has shaken up the profession. Its story is a master class in how to scale that "founder feeling" and create a workplace built on new power principles.

Buurtzorg began in 2006 with a vision of what nurses would *not* do. They would not be drowned in paperwork. They would not let strict procedure trump healthy culture. They would not be governed by out-of-touch paper-pushers at HQ.

Instead, they wanted to do the work that inspired them in the first place: caring for patients. They would connect with their community and make the decisions that matter *themselves*. They would decide when and how to visit people in their homes, dealing with everything from changing bandages to making end-of-life plans. They would be in charge of their own world.

This was not an original dream. Their list of hopes will be familiar to many who work in old power institutions and imagine a Jerry Maguire moment, when they finally say, "Enough!," stick it to the man, and blaze their own path forward.

But Buurtzorg had a plan to go with the fantasy. To get community health care right, it would put all the power in the hands of small, self-directed teams of ten to twelve nurses. Each team would take charge of the complete process of caring for their patients— and for the running of their own small enterprise. They would make their own decisions, minimize administrative oversight, and be connected to the other teams via a shared technology platform, allowing all the nurses in the network to learn and work together. They'd be free to experiment with new and creative ways to deliver care.

Ten years on, Buurtzorg has over 10,000 nurses in the Netherlands, operating in 850 teams, serving over 70,000 patients a year. To support all this, it has a back office of just 45 people. A KPMG study shows that its model delivers "a 50 percent reduction in hours of care, improved quality of care and raised work satisfaction for their employees." Its overhead is much lower than the average. So are its sick leave rates. So is staff turnover. And Buurtzorg has been repeatedly scored as having the highest employee satisfaction among all Dutch companies with more than 1,000 employees.

Its nurses are also achieving higher rates of patient satisfaction, even as they provide fewer care hours. With the patient at the center of their mission, and with the removal of the heavy hand of bureaucracy, they get better results in less time. Their patients get their autonomy back more quickly and have shorter stays in the hospital.

Contrast this with the traditional role of the nurse who, across the world, feels under-resourced and unable to spend sufficient time to support her patients. Research by the International Council of Nurses and Pfizer showed that aggression against nurses is widespread, with a quarter or more in almost every country reporting workplace bullying by colleagues. This culture is costly: a 2015 survey by the Royal College of Nurses showed that almost a third of UK nurses are seeking a new job.

What life is like at Buurtzorg: One founder's story

Madelon van Tilburg, a nurse for Buurtzorg based in Den Haag, isn't seeking a new job any time soon. But if you met her, you might be struck by a sudden wish to quit *your* job, move to Holland, and become a home-care nurse. She beams when she talks about her job, brimming with professional pride and purpose.

It hasn't always been this way. She used to work in a corporatized home-care organization that was obsessed with hierarchy and bureaucracy, targets and time slots. She was always rushing: clocking in and clocking out. She ended up being punished for spending too much time with her elderly patients, even when doing preventive care. "I had this meeting with the same manager every week and she would say, 'Oh, you work with Mrs. Brown. On Thursday you went over five minutes longer than was estimated. Why was that?'"

As she explains, life at Buurtzorg is very different. "When you start a new team, you have to find your own office . . . Setting up an office is the starting point of the team because you have to find your place in a community and find your way around the community . . . It's also very good for building up your team. You're working together, you decorate it together, and your team decides who's going to find chairs, who's going to paint the walls, and all that. It really feels like your own business . . . You do it together and then you have the common responsibility—that's very important."

This might seem like a needless distraction, something that would be more efficiently provided as a shared service by a specialist at a head office. But asking nurses to own the whole start-up process is a very intentional step. It is part of building a culture that feeds individual agency, and builds collective bonds among the nurses.

This culture continues in their daily operation. Like start-up founders, the nurses on Madelon's team find themselves doing a bit of everything, juggling nursing, making care plans, managing the office, hiring and firing, budgeting and connecting with the local doctors and hospitals: "Sometimes you find out about new talents of your colleagues or about yourself. Someone might say, 'Oh, no, I don't want to do the planning. I'm no good at it.' We'll say, 'Just try it,' and then maybe she'll find out that she really likes it and she's really good at it."

Every member of the team can freely view its budget and billable hours (with help from Buurtzorg's own "nurse-friendly" software, which makes this transparent across the whole network). No one can question the motives of the boss, because management in each team is a shared responsibility. The teams are deliberately small enough that consensus-based decision-making is workable.

The deep sense of ownership and collegiality Madelon feels produces the kind of job satisfaction that even the fanciest all-you-can-eat Google cafeteria lunch spread cannot buy. It also provides a foundation, based in trust, transparency, and mutual accountability, that enables her to focus on outcomes, rather than just manage workplace politics or hit corporate targets imposed from the top down. Like any good start-up founder, she sees her job as locating customer pain points and coming up with solutions in innovative ways.

What is particularly surprising is that Madelon has developed this sense of ownership while being part of a much bigger organization, which usually tends to kill the founder feeling. But for a large company, Buurtzorg is remarkably flat. In place of six layers of management occupied by people with titles like "Advice and Support Officer (Systems and Operations Management)," it employs just fifteen "coaches" across 850 teams to support and advise around any issues that come up. Coaches are typically more experienced nurses, not administrators or managers who have never been on the front lines. And they don't have management authority or hiring approval or veto rights (formally or informally)—they're more like mentors and advisers. The fact that each is assigned fifty or sixty teams acts as a check on coaches becoming managers. Other than the coaches, the teams are left to get on with it. Buurtzorg's motto sums it all up: "How do you manage professionals? You don't."

Madelon's most vital piece of infrastructure is her access to all the other nurses. BuurtzorgWeb is a kind of new power network that facilitates learning, peer support, and the development of shared resources across teams. It allows Madelon to confer with 10,000 other nurses if she has a question about a particular disease or a tricky patient situation. She can find a nurse with more expertise on a particular topic than she has, or someone in another region beyond the confines of her local team. Nurses share their knowledge by creating online courses for one another; Madelon has found the one on dementia especially helpful. This online community is much more vibrant than the usual corporate intranet because in a world without managers, peer support becomes key to professional development and problem solving.

In her old job, Madelon often used to hope that patients wouldn't ask her a question on the way out, knowing this would push her way over her allotted time quota. But at Buurtzorg, making time for a cup of coffee comes with the job. That can make all the difference, not just for her and her patient, but for health outcomes. Once people start talking, she finds, truths and solutions often emerge.

She proudly tells the story of one patient who started sharing her concerns about her ninety-six-year-old neighbor, also a patient of Madelon's. And she offered to help. So Madelon recruited her to stop by next door at the end of each day to help the older woman remove her support stockings. This not only saved Madelon the task, but also added a little more glue to her community.

What Buurtzorg gets right that Silicon Valley got wrong

An ocean and a continent away, another experiment with self-managing teams has taken place. Holacracy is a radical new management philosophy and system, invented by software engineers, that digital outfits like Zappos and Medium have been enthralled by in recent years. Zappos took it so much to heart that it restructured the entire company, eliminating bosses and departments altogether. On the surface, Holacracy appears very new power, offering a "new way of running an organization that removes power from a management hierarchy and distributes it across clear roles, which can then be executed autonomously, without a micromanaging boss." It boasts more flexible roles for staff and greater transparency.

Holacracy might sound a lot like Buurtzorg, only with better software. It is far from it. In fact, the differences between the two models help us understand the real secret to Buurtzorg's success, and what the rest of us might try to emulate. Nurses like Madelon and her teams have huge freedom to define their own working methods while Holacracy is *highly* structured and directive, perhaps more so than many old power organizations. A sample instruction from the forty-page Holacracy constitution reads: "As a Partner assigned to a Role, you have the authority to control and regulate each Domain of your Role. You may do this on a case-by-case basis when others request permission to impact one of your Domains, by considering the request and allowing or withholding permission."

Restrictions and protocols, with a good sprinkle of jargon, become the focus under Holacracy, not the work itself. Medium abandoned Holacracy after initially embracing it with gusto, saying it was "getting in the way of the work," and Holacracy is partly blamed for the departure of an unprecedented nearly one-third of the Zappos workforce in just one year, 2015, and for the company's being left off *Fortune* magazine's annual list of the best places to work for the first time in eight years.

Holacracy is like new power for robots. Julia Culen describes her experience when a long-standing Viennese-based consulting firm of which she was a managing partner adopted Holacracy: "It felt like being part of a code . . . an algorithm that is optimized for machines, but not for humans. Instead of feeling more whole, self-organized and more powerful, I felt trapped. The circles I was being part of did not feel empowering at all but [instead took away] my natural authenticity as well as my feeling of aliveness."

What Buurtzorg gets so right is it puts *human beings* front and center. It focuses on how to get teams of people to form deep connections with one another (which is why they turn nurses into interior designers). Buurtzorg uses technology to enable better peer coordination, but it doesn't lead with it, a lesson the Holacrats (and some in Silicon Valley) have yet to learn.

FROM "TILL DEATH DO US PART" TO "FRIENDS WITH BENEFITS": HOW TO COME TO GRIPS WITH THE TRANSIENT WORKER

"During the next year, if given the choice, one in four Millennials would quit his or her current employer to join a new organization or to do something different. That figure increases to 44 percent when the time frame is expanded to two years. By the end of 2020, two of every three respondents hope to have moved on, while only 16 percent of Millennials see themselves with their current employers a decade from now."—Deloitte 2016 Future of the Workforce, surveying people in twenty-nine countries

The CEO of a major company was recently doing a Q&A with a new class of interns. A feisty soul piped up. "Can you tell me about my opportunities for advancement?"

"Advancement?" thought the CEO. "We haven't even *hired* you yet."

When you understand the draw of the founder feeling, it's no surprise that many workers at big organizations are increasingly impatient to move up or move on. Instead of resisting this, LinkedIn's co-founder and chairman Reid Hoffman has embraced it. Everyone who interviews there is asked a version of a question that the upwardly mobile intern would surely appreciate: "What's the next job that you would like to have post-LinkedIn?"

For many old power thinkers, this is the professional equivalent of asking your brand-new fiancée what she thinks your second wife should look like. But Hoffman wants to put a stop to what he sees as the biggest lie the workplace tells us: *We are a family*. We shouldn't think of our workplace commitment as we do our marriage vows, "Till death do us part." Instead we should frame the employee-employer relationship as an "alliance," made up of short and clearly defined "tours of duty" that last two to four years. (With Chris Yeh and Ben Casnocha, Hoffman wrote a great book on these ideas called *The Alliance*.) Once each tour is complete, it will often be renewed, but sometimes it will see the employee move on, with both sides better for it.

It should come as no surprise that LinkedIn—the biggest career social network, arguably the platform that has done the most to collapse barriers for people wanting to improve their prospects and shift jobs—is trying to reimagine the social contract with its own employees. In doing so, it is addressing what is also a big challenge for the old power workplace: how to capture value from the many people who aren't looking to make a decades-long commitment to one employer.

Here are some of the clauses from Hoffman's sample "Statement of Alliance" which can be used by those embarking on a tour of duty. Consider how different this is from the "for better, for worse" promises we implicitly demand from our new hires in the old power world.

- I want you to help transform the company.
- In return, I and the company need to help you improve your market value and transform your career (preferably within this organization).
- While I am not making a commitment to offer lifetime employment, and you are not making a commitment to stay for your entire career, we will act to maintain a long-term alliance, even if the employment relationship ends.
- We the company expect this current tour of duty to encompass the time it takes for you to execute the following mission objective: _____
- I expect this tour of duty will last approximately the following amount of time: _____
- Here is what the results of a successful tour of duty look like for the company (product launches, process improvements, sales, etc.): _____
- Here is what the results of a successful tour of duty look like for you (knowledge, skills, accomplishments, recognition, etc.): _____

What's more, once your tour of duty is over, your relationship with the company is not. LinkedIn wants its alums to refer people it should hire, act as ambassadors, share ideas, and help shape and protect the company reputation.

Those organizations that want to build robust alumni networks will have to first make sure they are producing proud graduates whose loyalty will endure beyond their last paycheck. There is increasing risk today for organizations whose former employees are ready to rise up and speak out against them. One senior HR leader recently told us that the website Glassdoor—which displays anonymous (and sometimes scathing) company reviews from current and former employees—"keeps her up at night" and is now having a much greater influence on new hires than her company recruiting site ever did.

THE ATOMIZED AND UNCERTAIN FUTURE OF WORK

Not only are workers becoming more transient, work itself is becoming more atomized and impermanent. The contingent workforce is growing rapidly, thanks in part to the rise of "gig economy" platforms. In a world of fraying loyalties, tour-of-duty professionals, downsizing, wannabe founders, and hybrid musician/coder/furniture designers, up to 40 percent of the U.S. workforce can now be counted as contingent. Some of this shift can be explained by the cultural dynamics we have laid out in this chapter, but much of it is also driven by hard economic realities: there are fewer well-paid, secure, full-time jobs up for grabs.

The challenges of managing contingent and gig economy workers may be a glimpse into the future of management generally. If you're a company with a large share of contingent workers (or your business model is dependent on them), the tropes of traditional HR can seem rather quaint. The standard reviews and performance development plans that worked in the old power world don't make sense for a large, distributed, and contingent workforce.

Consider a platform like TaskRabbit, for which the vast majority of workers are contingent. Such businesses rely on an initial layer of what we can think of as *algorithmic management;* the very design of their platforms allows them to enforce rules and create incentives for desired behavior. Customer rating systems stand in for performance reviews. Allowing workers to rate customers can maintain morale by weeding out the bad apples; workers in traditional service roles rarely get to impose real consequences on a mean or haranguing customer.

But algorithmic management alone will only go so far. Creating a human connection among workers on a vast scale will also be critical. What a network like Buurtzorg is getting right is structuring for peer accountability and learning, which lightens the management burden and adds a powerful sense of shared endeavor.

Gig economy platforms are shaping expectations about work and working conditions, and mostly for the worse. In 2017, one such platform, Fiverr, which offers up freelance jobs and tasks that pay as little as $5, released an ad featuring a hip but exhausted-looking millennial and this wisdom: "You eat a coffee for lunch. You follow through on your follow through. Sleep deprivation is your drug of choice. You might be a doer." Fiverr says it champions "lean entrepreneurs" (quite literally, in the case of their advertising). The press release that accompanied the ad's launch read like a sad parody of the founder feeling: "The campaign positions Fiverr to seize today's emerging zeitgeist of entrepreneurial flexibility, rapid experimentation, and doing more with less. It pushes against bureaucratic overthinking, analysis-paralysis, and excessive whiteboarding."

In Fiverr's world, founders can look to some a little more like serfs. As Jia Tolentino observed in *The New Yorker*, "I'd guess that plenty of the people who advertise services on Fiverr would accept some 'whiteboarding' in exchange for employer-sponsored health insurance." Fiverr clearly fails the "circle test" we laid out in chapter 6.

Platforms can do better. Take the example of Care.com—the world's largest platform for caregiving. With more than 24 million members, Care.com is where millions of families in the United States and nineteen other countries go to find nannies, babysitters, and elder care. Its founder, chairwoman, and CEO, Sheila Lirio Marcelo, a Filipino-American entrepreneur, had the same early incentives to squeeze her caregivers that Uber had to squeeze its drivers, but she took a longer view. Marcelo herself was raised by Filipina nannies, and she understands the importance of their being treated with dignity and respect. As she told us: "If we want to sustain [this] for the long run you have to build a business that can't be driven off the backs of low-income labor . . . That won't be an attractive career for these people."

Instead, she chose to build a "people-to-people company," championing the rights of the caregivers on her platform. She has formed a deep and unlikely partnership with Ai-jen Poo and her National Domestic Workers Alliance, one that both sides were initially wary of. Together, Marcelo and Poo have advocated for improved treatment of domestic workers, including initiatives like the "Fair Care Pledge," through which families using the Care.com platform commit to fair pay and paid time off for anyone they employ. Plus, in an industry first, they have created a pooled, portable benefits program that allows workers who are paid via the platform to create their own social safety net. Now they are advocating on Capitol Hill and around the world for a comprehensive Care Workers' Bill of Rights.

As work becomes more tenuous, an organized voice for workers becomes even more critical. Unions continue to decline around the world, another twentieth-century form of affiliation that has struggled to adapt to the twenty-first. But new power models can help us reimagine them. A small glimpse of this is coworker.org, a platform that allows anyone to start a workplace-level campaign, with or without a union, from Starbucks baristas who agitated for more staff to Uber drivers who fought to have tips introduced on the platform, and won.

As we look to the future of work, it's easy to imagine a bifurcation: the vast majority of work shaped by the financial logic of automation and the mathematical logic of the algorithm, alongside a small number of hyper-empowered "founders" with tremendous agency, access to capital, and capacity to innovate. But we should reject the idea that our destiny is to end up either being replaced by robots or treated like them. We believe there's a practical advantage—and a moral imperative—to designing platforms that prove more human: ones that offer worker protections and security, provide freedom and dignity, and release creativity and ultimately value.

THE FUTURE: A FULL-STACK SOCIETY

Yochai Benkler is a professor at Harvard who has spent the past twenty years writing and thinking about the possibilities of technology-enabled collaboration and mass participation. He has, for most of his career, been an articulate cheerleader for the potential of the internet to distribute power—culturally, politically, and economically. His work has influenced our thinking greatly.

In the distant days of 2006, when Facebook was just a toddler, Benkler got into an argument that would grow into a big bet about the future (and would also sound a little like the title of a Robert Ludlum novel). It became known as the "Carr-Benkler wager."

It all began with some strong words. Nicholas Carr, the technology writer and commentator, posted a somewhat snarky response to Benkler's then-recent book, *The Wealth of Networks,* in which Benkler anticipated a wave of "nonmarket-based cooperation and productive collaboration," which translates roughly as "people creating things together for reasons other than financial reward." Wikipedia and open-source software are good examples. Benkler predicted that this kind of "peer production" was going to become a lot more common in a hyperconnected age.

Carr was having none of it. He described Benkler's book as a "techno-anarcho-utopian magnum opus," and you could almost taste his disdain in the hyphens. Instead of Benkler's peer-produced paradise, Carr anticipated a much different future. He saw this open-sourcing and collaboration, this hyper-friendly communalism, as part of a predictable and inevitable process of "amateur activity springing up in the wake of the invention of a new communication medium, only to be followed by increasing professionalization and commercialization."

The debate was a clash between Benkler's new power vision—that the huge opportunities in peer production afforded by connectivity would come to dominate the internet—and Carr's old power conviction that the forces of capitalism and managerialism would reassert themselves: collaborate vs. capture. So Benkler proposed a challenge to test their opposing ideas.

"We could decide to appoint between one and three people who, on some date certain . . . survey the web or blogosphere, and seek out the most influential sites in some major category: for example, relevance and filtration (like Digg); or visual images (like Flickr). And they will then decide whether they are peer production processes or whether they are price-incentivized systems."

Five years later, the moment of truth arrived. Who had won the Carr-Benkler wager? You may not be greatly surprised to hear that both men thought they had won.

In May 2012, Carr declared victory with the marvelously titled blog post: "Pay Up, Yochai Benkler!," underlining his view that "the dominant production systems in most online media categories are commercial ones." Benkler retaliated a week later with a post of his own: "For investors, no less than for social innovators and activists, the place to be is in sites that build platforms for social interaction and engagement, not places that figure out how to pay for the best content."

Carr fired back, citing music production and distribution, news reporting, video, e-books, games, apps, and many more as examples of the triumph of corporate consolidation over peer collaboration. He ceded photographs and encyclopedias to Benkler, and (probably wisely) demurred on pornography.

FROM PEER PRODUCTION TO PARTICIPATION FARMS

More than a decade after it was made, the wager remains unsettled.

They both turned out to be more than a little right. Carr was certainly correct that peer-to-peer activity online would become more commercialized and professionalized. The early flourish of something like Couchsurfing.com—an authentic, often charmingly ramshackle sharing platform for travelers in which no money changed hands—fell by the wayside with the rise of price-based options like Airbnb. But Benkler was ahead of his time in anticipating the huge (and unpaid) peer production that drives the models of everything from Facebook to activist movements.

Yet we can also see that the even unpaid peer-production models Benkler held up so hopefully have concentrated huge wealth and power in the hands of the owners of these platforms. The content we generate creates enormous value for powerful intermediaries that gobble up our data, opaquely decide what content we get to view, and claim the lion's share of the riches that result.

There are honorable exceptions like Wikipedia, but nonprofit, non-extractive models like this are not the norm. And in the meantime those paid models Carr anticipated that would reward our productive contributions are increasingly accused of using their scale and leverage to exploit those they pay for their efforts. Uber drivers are hardly being professionalized—indeed, Uber is doing everything it can to resist treating them as employees, as we've seen. In some senses, you might say the Carr-Benkler wager has turned out to be lose-lose.

When we paid a visit to Benkler at his office at Harvard Law School, he seemed a little glum. "I'm thrilled that you're writing this optimistic book in 2017," he told us. "I think it's genuinely harder to write such a book today than it was ten years ago."

To explain why, he describes a class he teaches every few years on the music industry and intellectual property. "What's interesting is that if you teach the case of the music industry in the late

2000s, what you see is radical decentralization and breakup. When you teach it five years ago, it's the beginning of the emergence of mid-level structures to allow artists to make a living. Voluntary donations, SoundCloud . . . Then you're teaching it in 2016 and it's basically all Spotify. It's completely *recentralized*." (You can imagine Carr rubbing his hands together in self-righteous glee.)

As consumers who are well trained to value the "frictionless" over almost all else, we all benefit from having easier-to-use services like Spotify (especially those of us who can remember failing to work out how to download obscure Tracy Chapman tracks on Napster). Yet the price of this is that we are all locked in to a handful of platforms that keep consolidating and making it harder for us to disentangle ourselves from them (just think of how Facebook eats up potential competitors like Instagram and WhatsApp so it can create one big ecosystem we depend on). And those creating the value on these platforms are squeezed. YouTube creators, Benkler reminds us, now have to sign over the rights to their catalog to the company in order to share in the revenues their content creates. The new power models that were a mere twinkle in Benkler's eye a decade ago have grown big and have become increasingly decoupled from the new power values their founders espoused. They are ever more useful in our daily lives, but they also feel and act more and more like participation farms.

IMAGINING A DIFFERENT WAY: CAMELS, NOT UNICORNS

Kickstarter is one of the original and best new power platforms. It has by now raised billions for creative projects, opening up huge new opportunities for artists and inventors. It has led a crowdfunding category that has significantly reduced the dependency of artists on funding from large traditional intermediaries like record labels. Unsurprisingly, Benkler and Carr have fought over what the success of Kickstarter proves. "Clearly it's about raising money to support artists, so it supports the interest in professionalism, à la

Nick," admitted Benkler. But he went on to argue that Kickstart-er's model of funding is closer to "Benkler's dream," in that "artists would get funded by people ponying up real money without expect-ing to make big bucks in return because of social motivations."

Kickstarter was funded by top venture capitalists like Union Square Ventures. It could easily have made Perry Chen, its first CEO, and his co-founders filthy rich—especially if it sought to fur-ther commercialize its core model. But Perry and Kickstarter took a very different turn. In a famous blog post in 2012, as the platform was reaching scale and becoming a cultural touchstone, the found-ers declared "Kickstarter is not a store," and introduced measures to steer the site away from simply becoming a way to preorder a product. Perry decided not to sell Kickstarter or take it public, but to turn it into a "public benefit corporation"—an emerging corpo-rate and legal structure that binds the company to serve society, not just its shareholders (building on the idea behind B Corporations [B Corps], a voluntary certification for socially responsible busi-nesses). Kickstarter's charter boldly, and in plain English, paints a picture of a very different kind of new power behemoth:

- Kickstarter will care for the health of its ecosystem and integrity of its systems.
- Kickstarter will never sell user data to third parties. It will zealously defend the privacy rights and personal data of the people who use its service, including in its dealings with government entities . . .
- Kickstarter will not cover every possible future contingency, or claim rights and powers just because it can or because doing so is industry standard.
- Kickstarter will not lobby or campaign for public policies unless they align with its mission and values, regardless of possible economic benefits to the company.
- Kickstarter will not use loopholes or other esoteric but legal tax management strategies to reduce its tax burden.

Perry told us about the conversation he had with prospective investors in the early days: "We told people, 'Hey, we're not building this thing to sell it. We're not building it to go public. We think those outcomes are likely to get in the way of actually having the most impact and we still think we can be a good investment . . . but we're going to do it in this way.'" We can think of Kickstarter less like the mythical "unicorn" that Silicon Valley seeks to capture—that exceptional but rare creature that delivers magical returns—and more like the commonplace camel, which is a little more plodding, but serves a critical function and, well cared for, can sustain investors and society for the long term. A study at the University of Pennsylvania found that Kickstarter projects have created 8,800 new companies and nonprofits, and a remarkable 29,600 full-time jobs. The platform has generated more than $5.3 billion in direct economic impact for those creators and their communities.

Perry thinks many more entrepreneurs can be moved to build camels, not unicorns: "Really it's to allow the generation that's coming up now, which I think has a lot of this inclination . . . to be able to know that there's not just one way to do it." Given rising expectations among millennials and those younger about the integration of profit and purpose, more Perrys are not only possible, they're likely. And the logic of his strategy is even more compelling for entrepreneurs creating new power communities—whose fate often rests on how well they cultivate and respect their crowds.

The journey—and philosophy—of Perry and his co-founders shows the path that might take us away from a life of participation farms, and toward one that is less extractive and more socially generative: platforms that meet the "circle test." Around the world, experiments are under way that take Perry's impulses much further and begin to reimagine new power models like Facebook, Twitter, or Uber in ways that more meaningfully distribute power and ownership.

The University of Colorado, Boulder's Nathan Schneider is one of the leaders of a growing (but still quite academic) movement that is championing what he calls "platform co-ops," democratically run and governed cooperatives reimagined for a world of peer-based technology platforms, not just farms and factories. This movement wants to turn the participation farm into something that looks more like a digital kibbutz.

Schneider points to models in more traditional industries that run along these lines, like the UK department store chain John Lewis, which in 1929 was put into a trust owned by its employees, who share in the retail chain's profits and elect representatives to its governing board. For new power platforms, he argues, it is their millions of users, not just those on the payroll, who should share in the value created, have a say in big decisions, and be represented in the governance of these platforms. He has proposed that Twitter's users try to buy it back, arguing that it serves an essential public function. In his mind, for all its challenges, the problem isn't that Twitter isn't working for its users—he cites the powerful justice movements that rely on it as evidence that it is. The problem is that "Wall Street's economy has become Twitter's economy." Put a little differently: in Carr's terms Twitter is underperforming; in Benkler's it is thriving.

The #BuyTwitter movement championed by Schneider and others was significant enough that it ended up as one of the five proposals on the table at Twitter's 2017 annual general meeting. It made a strong case for a Twitter that would function very differently.

A community-owned Twitter could result in new and reliable revenue streams, since we, as users, could buy in as co-owners, with a stake in the platform's success. Without the short-term pressure of the stock markets, we can realize Twitter's potential value, which the current business model has struggled to do for many years. We could set more transparent accountable rules for handling abuse. We could reopen the platform's data to spur innovation. Overall, we'd all be invested in Twitter's success and sustainability. Such a conversion could also ensure a fairer return for the company's existing investors than other options.

This motion was not embraced by Twitter Inc. And it would be very tough to flip Twitter into a co-op in this way. But it points to a compelling alternative vision, for us as participants and for the next generation of platform creators.

In fact, companies with this cooperative-inspired philosophy and model are beginning to emerge. The photo-sharing co-op Stocksy brings together photographers and filmmakers, giving them an opportunity to license their work. They are a proud platform co-op but also a serious and growing multimillion-dollar business. As they put it: "(Think more artist respect and support, less patchouli.) We believe in creative integrity, fair profit sharing, and co-ownership, with every voice being heard."

For platform co-ops and similar ideas to succeed, governments will have to make it easier to raise money to scale without relying on big investors or the traditional capital markets (if that had been the case, Perry might never have needed to raise venture capital for Kickstarter in the first place). That means promoting large-scale crowd equity of the kind that BrewDog has relied on—with government playing a role to ensure real accountability and transparency so small investors aren't taken advantage of.

To truly reimagine the platform behemoths, we also need to reimagine their algorithms. As Facebook has shown us, social media sites have huge power to alter our consumer preferences, spur or hinder extremism, and sway our emotions with tweaks of code. But today their algorithms function as secret recipes that serve private interests.

Consider how a "public interest algorithm" might work instead. What might a formula designed to favor the interests of platform participants and society at large—instead of just their owners, advertisers, and investors—look like?

It would need three key features. First, the inputs into the algorithm, which shape what content we see and what gets priority, would be fully transparent to the user, including the criteria used by the platform to moderate offensive content or hate speech. Second, every user would have a range of dials that allowed them to alter their world. They could choose to engage with more content they disagreed with. They could "filter in" perspectives and views from those well outside their bubbles. They could reduce sensationalism. Third, the default settings of the algorithm would apply a public interest test, considering how the platform can better serve its broader "circle." This might operate like an updated version of public broadcasting, bringing to the surface content proven to reduce social tension and extremism and bolster civic discourse, promoting pluralism, and showcasing unserved and underserved communities. Participants and super-participants in the platform, and the wider public, could have legitimate debates about whether, how, and how much a platform should "tip the scales" in this way.

NEW POWER, NEW PLATFORMS, NO PLATFORMS?

A decade ago, none other than the father of the World Wide Web, Tim Berners-Lee, saw the dangers of participation farms on the horizon.

In 2008, almost twenty years after laying out his original vision, he rallied for the building of "decentralized social networks" that would reclaim his beloved web from increasingly centralizing sites like Facebook. He saw a big prize in a more fluid and pluralistic world of platforms in which "online social networking will be more immune to censorship, monopoly, regulation, and other exercise of central authority."

Today, he is hard at work on a project to address that very issue, a plan to radically alter the way web applications work, one that would divorce all our personal data and content from the apps and platforms that now—often literally—own it. Berners-Lee's Solid project would allow us to own our own data as part of a personal secure "pod" in which we would carry around our digital lives. So imagine that, rather than having all your data on a third-party platform, you now take it with you. (This is what geeks call "interoperability.") You walk around with your photos, friends, health histories, a map of all the places you have traveled, a list of all your purchases—even the online reputation you have built up in various platforms, an especially powerful commodity. You are liberated to decide what access you would like to grant—and on what terms—to whom. Solid is much more than a different kind of technology; it is a different philosophy. With Solid your data "reports to you."

Another solution to the problem of placing one's data in the hands of a powerful intermediary comes in the great—and much-hyped—hope of the Blockchain. The Blockchain is a distributed public ledger that allows everyone to record and see what transactions have taken place. Unlike a centralized secret ledger—such as those of banks—it is transparent. And transactions are verified not by a central force, but as a distributed process. You might know the Blockchain from its most famous (and controversial) application to date: it is the underlying technology upon which the virtual currency Bitcoin is built on.

For non-technologists—even those who have spent hours trying to get their heads around this—the way this *actually* works can be hard to grasp. But the most important things to understand are the potential human applications. As *The Economist* puts it, "It offers a way for people who do not know or trust each other to create a record of who owns what that will compel the assent of everyone concerned. It is a way of making and preserving truths."

The potential of this is as huge as the hype (although, like all technologies, Blockchain remains vulnerable to co-optation and capture). It opens up a world where users might exchange value directly without an extractive middleman. We can easily imagine real estate contracts or financial transactions living on the Blockchain. But we might imagine, too, the intermediaries being removed from the mega-platforms of the world—our Ubers or Airbnbs—when drivers and riders, or hosts and guests, work out ways to collaborate and exchange directly with each other.

As we look to the future, there is no shortage of predictions about the next participatory technologies and ideas that will transform our lives. Whether it be virtual reality, augmented reality, blockchains, or even the emergence of a metaverse, platforms as we know them today will likely end up feeling rather quaint. But however things turn out, we need to cling to, and build for, a set of principles that ensure the worlds we will live in are less monopolistic, more transparent, and much more attuned to their broader impact.

We, as the participants on the participation farms, can do more than just lament our fates. In a *Guardian* poll in 2017, less than one-third of Americans agreed that Facebook was good for the world, and a paltry 26 percent believed that Facebook cared about its users. The poll also showed support for placing limits on the power of the big technology platforms. But there is no reason to leave this work to regulators, who will tend to be more focused on blunting the power of platforms than on preserving their benefits to users. Consider some of the stories we have shared in this book and you can begin to glimpse a movement to flee these platforms that takes on some of the viral and participatory qualities of an Ice Bucket Challenge. And while no movement is likely to empty out the farms altogether, even a relatively modest user rebellion might trigger big changes to policies or leadership, as we saw in the wake of the #DeleteUber effort in early 2017. It will take inspired and dedicated applications of new power—by technologists, entrepreneurs, and all of us—to reinvent the terms on which we participate.

TOWARD A "FULL-STACK SOCIETY"

The political thinker and social innovator Geoff Mulgan describes one of the great paradoxes of our age: "People were promised that the currents of change, economic, social and technological, would make them feel powerful. Instead, they see decisions being made by political and corporate leaders ever further away from them. They feel like observers, not participants." He cites research that shows that majorities of people across the world think their countries are on the wrong track.

We saw this paradox at work in the Brexit vote. The "visceral" difference in the minds of "leave" and "remain" voters, in the view of the British Election Study, was a "deeper sense of alienation." Those who wanted out tended to feel less control over their own lives. In the same vein, Nobel Prize–winning economist Angus Deaton showed a clear correlation between predictors of what he called "deaths of despair"—from drugs, alcohol, and suicide—and support for Donald Trump.

The fear of automation and the stark realities of income stagnation and rising inequality are all feeding this sense of alienation. Even the so-called "winners" of globalization feel distant from core social institutions, from government to media to education, as the Edelman Trust Barometer shows. We can vent our frustrations in an instant on social media and find many who will cheer us on. But we may still feel less control over the hard realities of our lives, how we fit into society, and how we engage with our governments and institutions.

To close this gap, it will be critical to actually reduce wealth and income inequality and change the material conditions of those who have been left behind. But a subtler challenge is in how we create more meaningful opportunities for people to actively shape their lives and connect with the institutions that shape them. People need to feel more like owners of their own destinies, rather than pawns of elites. If the only meaningful expression of all this pent-up agency is the occasional election or referendum, people will naturally be inclined to use their participation as a way to lash out. Platform strongmen and extremists will offer easy answers. But we need something different: a world where our participation is deep, constant, and multi-layered, not shallow and intermittent.

Think of this world as the "full-stack society."

When coders speak of stacks, they are referring to the different components of software that work together to make a product hum: programming languages, applications, and an operating system. The "full stack" is when all of those layers come together to form a coherent whole, both those that we see (the "front end") and those that we don't (the "back end").

This is a good analogy for the kind of world we need to create, one where people can more meaningfully participate in and feel ownership over every aspect of their lives—their engagement with technology platforms, their work, health and education, and of course democracy and the experience of government itself. This cannot come about through one "killer app," but requires a cultural and structural shift, with institutions of all kinds building deeper and more rewarding routes to participation. The most important reason that people don't trust institutions is that institutions don't really trust people, offering occasional, insubstantial, and often frustrating opportunities to take part. To create the full-stack society, we need to dream up entirely new models that make us feel more powerful and more connected to one another in all our guises: as patients, taxpayers, consumers, neighbors, voters, students, and parents.

BUILDING UP THE STACK

To help us imagine what a full-stack society might look like, here are two stories of people adding layers to the stack in critical areas, one in media, one in government. Each teaches us something about how to make people feel more powerful, and how to make society stronger. And they offer a critical lesson: if we want to rebuild public trust in our institutions, we have to let people get their hands on them.

From subscribers to protagonists:
The story of *De Correspondent*

In 2009, Rob Wijnberg, while still in his twenties, became the youngest ever editor in chief of the prestigious Dutch newspaper *NRC Next*. He was hailed as a boy wonder, and already had a best-selling book to his name. A philosopher by training, and with news in his blood (both his parents are journalists), he had built a reputation as thoughtful and earnest, a throwback to that rose-tinted age of newsrooms brimming with reporters focused on the public interest rather than sponsored content.

Two years later he was fired. He had tried to move the paper away from breaking-news-driven reporting, which he saw as prone to hype, sensationalism, and a short lifespan, in favor of more analysis, coverage of structural issues, and long-form investigative journalism. It is salacious and exciting to ogle the various billionaires caught up in the Panama Papers. But, he told us, it is much more important that readers understand the underlying global system of tax evasion and capital flight. This unusual perspective didn't jibe so well with the business goals of *NRC Next*'s owners, and when he refused to change his approach, he got the boot.

This used to be the point where the story ended. In the twentieth century starting a newspaper was a vastly expensive endeavor requiring all manner of physical infrastructure and capital costs, which is why fired editors didn't usually start new ones. But Wijnberg was not one to give up. If he couldn't rethink journalism within an old power institution, he'd do it by creating a new power one.

Wijnberg laid out a vision for a new kind of publication that would retain the essentials of high-quality journalism while rethinking just about everything else. This new publication would reinvest at least 95 percent of its revenue back into journalism and the development of its platform. Its privacy policy would explicitly prohibit it from collecting unnecessary personal information about its readers. It would be entirely free of ads. It would be "openly subjective," rejecting the myth that journalists were "objective automatons," and encourage reporters to actively pursue real-world change. And, most central, it would reimagine the relationship between a publication and its readers by turning them into active participants in the journalistic process.

He took his pitch to the crowd, inviting people to step up to support his vision and become his very first subscribers. He broke world records for crowd-funded journalism. His creation, *De Correspondent*, raised 1.7 million euros in thirty days, and on day one the publication already had 19,000 subscribers (called "members").

At the core of *De Correspondent*'s proposition is a critical new power idea. As Ernst-Jan Pfauth, Wijnberg's co-founder and publisher, puts it: "The people previously known as 'the audience' constitute the greatest source of knowledge, expertise and experience to which journalists have access. Yet this resource has been underutilized for more than a century and a half." He adds: "We believe that modern journalists shouldn't see their readers as a passive group of annoying followers. Instead, they should regard readers as a potential gold mine of expert information."

This comes to life in many ways at *De Correspondent*. Every journalist shares their story idea or the question they want to investigate with members before they write a single word. Readers are then invited to draw on their expertise in that topic and share insights, ideas, or leads. Readers are given tools to curate, build on, and adapt material from the newspaper, adding external sources and their own experiences. And every article in the newspaper ends with a question for readers that "steers the contributions in the direction the correspondent finds journalistically most relevant."

De Correspondent reframes user "comments" as "contributions" and has turned them into an integral part of the experience, not an afterthought. Editors and journalists invite specific members to write guest posts in response to articles they have engaged with. They treat readers like experts by allowing them to add a tag to their names reflecting their areas of special knowledge and to add their own bio (unheard of on traditional media sites, where the only bio you'll see is of the writer).

In doing all this, *De Correspondent* sees itself as cultivating "the world's greatest rolodex"—a trusted community of expert contributors, from scientists to nurses, who help its journalists write better stories and gain a more diverse range of sources and interlocutors.

Their readers love to get in on the action. *De Correspondent* is currently conducting "the Netherlands' largest group interview with refugees," asking readers to pair up with refugee asylum-seekers, meet monthly to interview them, and share their (often ignored) stories. Hundreds have taken part.

To take another example, when environment correspondent Jelmer Mommers wanted to launch an investigation into Shell, he started with a blog post that asked: *"Dear Shell employees: let's talk."* Rather than hoarding his findings until the "big reveal" of an article, he shared his growing knowledge and questions with readers and sought their help. "The updates Jelmer published amplified his reach and trust. They brought in new readers and new sources," Ernst-Jan Pfauth explains. One of those readers handed him a treasure trove of documents that proved Shell understood the grave risks posed by climate change decades before it admitted it.

A handful of years after its initial crowdfunding campaign, *De Correspondent* is thriving with more than 52,000 subscribers (on a per capita basis, not far off that of major publications like the *Wall Street Journal*, and it is still mainly in the Dutch language). It has retention rates that most media companies would kill for: 79 percent on its annual subscription, and 89 percent on its monthly plan. And it is now planning to take its model global.

In an era of deep skepticism of traditional media, incessant talk of "fake news," and a U.S. president who has described the media as "the enemy of the people," *De Correspondent* shows us how to bring participation and a sense of shared ownership back into a vital social function. Those who have lost trust in media are offered an opportunity to get their hands on it, not as passive readers, but as real participants. In doing so, people become part of the adventure. Newspapers like *The Guardian* and the *New York Times* are beginning to cotton to this, launching big "membership drives" that emphasize their higher social mission. But the real participation premium will be won by getting off their pedestals and inviting their readers to join in.

Forking government: The tale of Audrey Tang

In Taiwan, a brilliant and fearless thirty-six-year-old transgender coder and activist, Audrey Tang, known as one of her country's best technologists, is showing what is possible when new power is brought into the heart of government.

Her journey began in 2012, when the Taiwanese government

released an ad depicting citizens gazing awestruck at the complex details of an economic reform package that appeared above them. "The message was essentially: it's too complicated for you, but do not worry, trust us. It was really insulting," Audrey told Claire Richard at the *Open Culture* blog. So a group of hackers made a website that transparently published all government budget data and made it easy for citizens to engage and debate the facts. They called it g0v .tw, a clever "forking" of an official government website address, by swapping out the letter *O* for a zero.

The key technical challenge was to create visualizations compelling enough that anyone could understand what the government was actually up to. To do that, you have to scan a lot of data. The "g0v" wanted to enlist the crowd, but they had to find a technical solution for how to get people involved: "If we had asked people to copy an entire page of numbers in an Excel table, that would have taken maybe four minutes," said Tang. "On the Internet, it is already too long: people can find instant gratification elsewhere, they can simply 'like' or 'share' on Facebook in seconds. It should not take more than a minute: it is the limit on the Internet."

So they cut the vast trove of data on government spending into "captchas" that take just a few seconds to transcribe, and added badges to motivate people along the way. As Tang put it with characteristic enthusiasm: "It takes five seconds and it really feels like you're helping the country! This is both useful, simple, rewarding and fun. This is the key of crowdsourcing: whenever there is a way to measure the progress of the player, people can spend hours without sleep to finish the game!"

The project attracted 9,000 people within twenty-four hours.

Tang and g0v's next trick, in 2016, was to coordinate among the traditional taxi industry, the Taiwanese government, and ridesharing companies like Uber to figure out a novel approach to regulation. She used an online platform called Pol.is to orchestrate a remarkably nuanced consensus-building process that brought 4,500 users and drivers together to propose solutions and live-streamed the decision-making. A new law emerged that extracted major concessions from Uber, including requiring Uber to vet drivers more fully and provide insurance, and that made it easier for civil society alternatives to Uber to spring up.

In her journey, Tang went from being an activist operating outside the system to becoming a visionary politician and the youngest minister without portfolio in Taiwan's history. In office, Tang has hired what she calls "Participation Officers," whose job is to create mechanisms for meaningful participation in the work of government by ordinary people, something that seems obvious but somehow still feels deeply unintuitive for the twentieth-century managerial state.

Tang is determined to find ever more ways to get people participating (and co-owning) the outcomes of government. Platform co-op acolyte Nathan Schneider is on the same push: he believes the key to moderating and deepening our public discourse is for people to be constantly debating, voting, deliberating. He calls this "unbundling democracy": "I believe in the idea of unbundling our democracy so that you're not just electing one representative to represent you on everything, but to imagine democracy as a series of layered and cascading institutions that you relate to according to your relationship with that service." He imagines this applying to everything from electing the governors of a health-care system to choosing who runs the big online platforms in our lives.

To play this out a bit, think for a moment of the humble (and hated) annual tax return. Imagine if it were no longer impenetrable and dull, but instead fully immersive and participatory, with every citizen hearing the stories of those people their taxes had helped and seeing where their money had gone. Instead of being a dreaded once-a-year moment, it would provide feedback to citizens throughout the year. People might be surprised by a thank-you video out of the blue from a citizen whose life their tax dollars had changed: the worker in a declining industry who had been retrained, the kids enjoying a new playground, the doctor who had developed a new cancer treatment. People might even get to play a more active role in deciding where a portion of their tax dollars go.

Some of this is already happening, especially at the city level. Participatory budgeting is an idea that started decades ago in the

Brazilian city of Porto Alegre but is now spreading throughout the world and merging with internet crowdsourcing culture. Paris's ambitious participatory budgeting program, led by Mayor Anne Hidalgo, has led to hundreds of thousands of Parisians passionately debating and voting on projects close to their daily lives, everything from greening public space to aiding the homeless. This mirrors the success of the citizen engagement and budgeting program in Reykjavik, which has attracted nearly 60 percent of that city's people.

Plans like this must account for the bias inherent in any crowdsourcing exercise—especially the challenge of distinguishing between what is vital and what has viral appeal, and ensuring that they don't mostly benefit the already privileged. But done well, they can serve a larger purpose: giving people a sense of ownership and many more opportunities to participate.

These are just a few examples of the kinds of efforts we need in order to build up the stack. It is easy, and exciting, to imagine more. The health sector, for one, is made for more participation—and in an aging society it will be a critical place to focus. In fact, greater levels of patient activation have been called the "Blockbuster drug of the 21st century."

National Health Service innovation whiz Helen Bevan asks us to consider caring for our health as a contribution to our democracy: an ongoing investment in the personal and public good. To encourage this, the health layer of the stack will need to be built out to enable vibrant patient communities in the style of PatientsLikeMe, the platform we described at the very beginning of this book. We also see promise in the emergence of data and cloud co-ops for medical histories—places where people can store their personal information out of the prying hands of big platforms and then "peer-produce" better health outcomes.

More routes to deeper participation in domains typically closed off to the public will be key. Take the volunteer emergency services network GoodSAM, which started in the UK, allowing ordinary people with first aid training to support and augment big health systems. In the words of Alex Grant, a nineteen-year-old student paramedic:

> As a GoodSAM Responder, I volunteer my lifesaving first aid skills in the local community around me. During one partic-ular GoodSAM alert, I responded to a patient a few hundred yards from where I was who had suffered a cardiac arrest. We were quickly backed up by the London Ambulance Service and worked tirelessly for 40 minutes to save this person's life. We managed to get a return of spontaneous circulation. The patient sustained their pulse to hospital—long enough for the family to make the 200-mile journey where they could say their goodbyes and hold hands for one final time. Not all stories have a happy ending, but you can certainly try to make a difference—through comfort, sympathy and provid-ing a fighting chance for survival.

GoodSAM is now big enough that it can be a major complement to ambulance services in UK cities, but it has another advantage, too: it helps bring communities closer together and gives people like Alex Grant a deep sense of participation.

Across sectors, you can start to imagine initiatives of all shapes and sizes. Think of new power platforms that make it easy for retir-ees to teach and mentor the unemployed in an age of increasing automation; huge ecosystems of local, face-to-face groups in which people of different backgrounds can form communities focused on anything from learning Hindi to fighting racism. Think of the com-ing distributed, clean-energy future that promises to allow ordinary people to capture, use, and even sell power back into the grid and to one another.

We all need to build models—and seek out models—that help everyday people like us gain a sense of ownership and connection to one another and to society at large. "Participation" needs to be much more than the website that allows you to point out occasional potholes in the street; it needs to be a constant and compelling expe-rience that keeps people working together on the things that matter.

Building the full-stack society will be challenging at a time when public support for democracy itself is at historic lows in much of the industrialized world and especially among young people. And there will always be tension between getting people to show up for things that really count and the easy temptation for all of us to plug into the dopamine machine. It will require that those on the side of the angels get a lot better at delivering a killer user experience, sticky feedback loops, and a compelling set of incentives as they work to restore vitality to our essential social functions.

NEW POWER'S MESSY MASTERPIECE

On Reddit, over seventy-two hours in April 2017, something strange but unexpectedly beautiful happened. It's a tale that in many ways sums up the big unresolved question as we end this book: whether new power will ultimately do more to bring us together and build a more just world than it does to divide us and exacerbate inequalities. Reddit put a simple challenge to its community: it created a vast, blank virtual canvas it called "Place" in which users were invited to contribute, pixel by pixel (with a limit of one per user every five minutes), to making one big piece of art. The dominant early contributions were penises, swastikas, and swear words (if you recall the tenor of #RedditRevolt back in chapter 5 this won't much surprise you). But soon, the better angels on Reddit responded with pixels of their own.

A product manager at Reddit remarked that "what was really amazing was seeing how quickly the community organized and started to self-police the canvas to keep it positive." A stunning replica of Van Gogh's "Starry Night" appeared and, thanks to the help of others in the community, survived an attack from a nihilistic group of black-pixel brandishing contributors who called themselves "The Black Void." National flags were drawn; countries fought and then made peace. Adult Fans of Lego inserted themselves to promote their beloved bricks. A man wanted to use Place to make a marriage proposal and enlisted the help of the community to make it look beautiful and keep it there. His plan was cruelly sabotaged but then partially saved by a community that devoted itself to painting hearts. More than a million people laid out 16 million pixels to create a complex, chaotic collaborative masterpiece, by some measures the biggest piece of group art ever.

Ultimately, the forces of light outnumbered and out-organized the forces of darkness on Reddit Place—and they did it by embracing collective action and through the strength of their core values. Reddit Place became a metaphor for what a "full-stack society" might look like—with many expressions of political, economic, and cultural participation represented on one big digital canvas. Everyone who contributed felt like an owner because each of those million people really was one. Building this kind of world will be messy—and it certainly won't be without pockets of hate, co-optation, and a whole lot of distraction. But it is a world all of us can create together, and it is the one most worth fighting for.

GLOSSARY
How to Speak New Power

An ACE idea: An idea designed so that the crowd will take hold of it and spread it. It is *actionable* because it is designed to make a user do something, *connected* because it makes a user feel part of a like-minded community, and *extensible* because it is structured with a common stem that encourages its communities to alter and extend it.

Blend power: To bring together old and new power in ways that reinforce each other.

Bridge: A new power change agent who can meaningfully connect with the new power world, making the practical "jumps" between old and new power. A bridge's work is structural. A *digital beard* is often mistaken for a bridge, but actually provides cover for a risk-averse organization and resides on the margins of influence.

Camel: A new power platform or organization that delivers economic returns that are less spectacular than those of its alter-ego the "unicorn," but serves an important social function and can sustain investors and its community for the long term.

The circle test: Asks whether the impacts of a new power platform, on those both inside its triangle and within its wider circle, are positive.

Connected connectors: People who are connected to one another, share a worldview, and are influential in their reach. For a new power movement, identifying and cultivating the right connected connectors can be the difference between takeoff and fizzle.

Crowd-jacking: When a campaign is separated from its original intent through the whimsy of the crowd. (Take the rise of Boaty McBoatface.)

Founder feeling: The increased sense of agency, control, and ownership employees seek, as if they were start-up founders, even if they work for someone else.

Full-stack society: A world in which all major social and economic institutions are designed so that people can more meaningfully shape every aspect of their lives.

Harnessing the three storms: Successful movements often build up off the backs of **storms**—galvanizing moments of human drama and urgency that can be hard to predict but that provide great energy. Sometimes storms happen and the movement works to **embrace** them, even if at first they seem like setbacks. Sometimes, a movement sees a storm out in the world and **chases** it. At other times, a movement **creates** a storm out of thin air.

Leadership compass: Describes leaders in terms of their leadership models and values. The *Crowd Leader* combines a new power leadership model with a commitment to new power values. She wants to do more than channel the power of her crowd, she wants to make her crowd more powerful. The *Cheerleader* champions new power values like collaboration, transparency, and participation, but leads in an old power way. The *Co-opter* deploys a crowd and skillfully uses new power tools and tactics, but does so in service of old power values, and to concentrate power for himself. The *Castle* pairs old power values with an old power leadership model. This is the traditional hierarchical and authority-based model of leadership most of us grew up with, and which is widespread in sectors like the military, business, and education.

Meme drop: A message or idea designed to spread sideways, coming most alive when remixed, shared, and customized by peer communities. Examples: the Ice Bucket Challenge, Pepe the Frog.

New power: Made by many, it is open, participatory, and peer-driven. It operates like a current and, like water or electricity, it's most forceful when it surges. The goal with new power is to channel it.

Old power is held by few. It is closed, inaccessible, and leader-driven. It operates like a currency. Once gained, it is hoarded and the powerful have a lot of it to spend.

New power behaviors: Sharing, affiliating, funding, producing, and shaping.

The *old power behaviors* are complying and consuming.

New power brand: A brand designed to inspire participation.

New power circle: The group of interested parties who surround a new power community and can impact its fate and are impacted by it.

New power compass: Plots organizations along two dimensions: models and values. A *Crowd* is a new power model with new power values. A *Cheerleader* professes new power values but sticks to an old power model. A *Castle* is an old power model with old power values. A *Co-opter* is a new power model used to reinforce old power values.

New power decision tree: Four questions to ask to decide if your organization should turn to new power. They address strategy, control, legitimacy, and long-term commitment.

New power models: Enabled by the activity of the crowd—without whom these models are just empty vessels.

Old power models are enabled by what people or organizations own, know, or control that nobody else does—once old power models lose that, they lose their advantage.

New power triangle: The three components of a new power community, including participants, super-participants, and the owner or stewards of the platform.

New power values: A belief in informal governance, opt-in decision-making, collaboration, radical transparency, maker culture, and short-term affiliation.

Old power values include a belief in formal governance, competition, confidentiality, expertise, and long-term affiliation.

"Occupy yourself": The technique of embracing radical transparency and surfacing your challenging issues, if only as a preemptive strategy. A key task for many old power institutions.

Participants: Those who take part in a new power platform, and tend to form the vast majority of its users.

Participation farms: Digital platforms that have fenced in, and harvest for their own gain, the daily activities of millions or billions of participants.

Participation premium: The additional value created by bringing together a higher purpose, a material return, *and* participation within a transaction.

The participation scale: Lays out the behaviors that old and new power models are designed to inspire, from least to most participatory. The most effective new power models know how to move people "up the participation scale."

Platform owners: The entity that controls who is allowed to participate in the platform and its governance, deciding how value is distributed and even whether the platform lives or dies. Where there is no owner, *platform stewards* play sometimes informal leadership roles.

Platform strongman: A Co-opter who deploys new power tools and methods in order to advance authoritarian values.

Problem solvers and solution seekers: Terms surfaced by NYU professor Hila Lifshitz-Assaf to describe, respectively, those within an organization who resist efforts to open up to the participation of the wider community and those who more easily make the shift. *Problem solvers* find their identity in the old power value of expertise while *solution seekers* expand the boundaries of their professional world to invite outsiders in.

Shapeshifter: A new power change agent in old power garb, a figure with unimpeachable institutional credibility who smooths the path to change. The shapeshifter can be contrasted with the *disrupter,* the outsider who "moves fast and breaks shit."

Shaping: How a new power leader influences the overall norms and direction of her crowd, especially in ways that go beyond her formal authority.

Signaling: The way a new power leader makes a crowd feel more powerful through his speech, gestures, or actions.

Structuring: Putting structures and practices in place that enable participation and agency. A new power leader "structures for participation."

Super-participants: The most active contributors to a platform, and often those who create the core assets that power the platform and create its value.

Taking the turn: To move from old to new power.

"WeWashing": A phrase coined by our friend Lee-Sean Huang to describe the way brands sometimes use the language of the crowd and appear to invite its participation without having any meaningful interest in engaging with it.

ACKNOWLEDGMENTS

Henry and Jeremy want to thank:

Our editor extraordinaire, Kris Puopolo, who acted as the best traveling companion, navigator, and occasional taskmaster any first-time authors could wish for. At Knopf Doubleday we owe much to the leadership of Bill Thomas, who backed this book with such enthusiasm, and to Markus Dohle and Sonny Mehta, who early on saw the book's potential. Our team there—Michael Goldsmith, Suzanne Herz, Daniel Meyer, and John Pitts—has been a delight to work with and added so much value to this process.

Our amazing agent, Elyse Cheney, who believed in us from the start and offered course-correction, good humor, and sage advice all in perfect measure. Alex Jacobs, Alice Whitman, and everyone else at the agency were a joy from start to finish.

Those who offered detailed feedback and insight on the finished draft: David McCann, Rob Reich, and Ali Weiner.

Our wonderful researchers: special thanks go to Campbell Schnebly, Kait Sheridan, and Allie Tempus, who contributed great diligence and energy. We are also grateful to Jon Alexander, Arthur De Grave, Irenie Ekkeshis, Vicky Ge Huang, Lee-Sean Huang, Antonin Léonard, and Lauren Rhue. Our fact-checker, Julie Tate, was terrific.

Melissa Cronin, who shepherded the book process and did a

superb job of engaging a community of people interested in these ideas. Robbie Ross, who played a peerless role throughout this project and continues to guide the new power world with deft skill. Jenn Manzetti and Annie Melia for keeping the new power show on the road and doing so with great patience, professionalism, and passion.

The teams at HBR and TED—especially Amy Bernstein, Bruno Giussani, and Adi Ignatius, who helped introduce our thinking to the world in 2014. We were excited to see HBR China and TEDx Amsterdam then take the baton by creating events dedicated to expanding on the thinking.

The new power practitioners and thinkers who are working to break new ground and who offered us their partnership and insights along the way: the NHS Horizons team under the remarkable Helen Bevan, Jack DeGioia at Georgetown, Charlene Lake at AT&T, and Jennifer Sampson at United Way.

We are also grateful to David Aikman, Danah Boyd, KJ Boyle, Richard Branson, Victòria Alsina Burgués, Paul Burke, Dan Cardinali, Beth Comstock, Michael Faye, Primavera de Filippi, Brian Forde, Kate Garvey, Phil Griffin, Cecilie Heuch, Jeff Klein, Karim Lakhani, Ellen McGirt, Beth Noveck, Alessandra Orofino, Shaifali Puri, Aaron Sherinian, Richard Socarides, Arun Sundararajan, Paul Tagliabue, Art Taylor, Darren Walker, and Kyle Westaway.

Special thanks to all who shared their insight and knowledge: Mayor Steve Adler, Yun Mi Antorini, Phil Aroneanu, Hila Lifshitz-Assaf, Douglas Atkin, Ben Balter, Zachary Barrows, Yochai Benkler, Nadia Bolz-Weber, Randy Bretz, Letitia Browne-James, Craig Calhoun, Harry Campbell, Jennifer Carlson, Alexandra Cavoulacos, Perry Chen, Joe Deshotel, Alex Fiechter, Aria Finger, Natalie Foster, Ligia Friedman, Chuck Gates, Mark Glaze, Reuven Gorsht, Nicola Greco, Tate Hausman, Scott Heiferman, Ahti Heinla, Naomi Hirabayashi, Philip K. Howard, Rick Ifland, Verity Jones, Ben Keesey, Jess Kutch, Joseph Kvedar, Sheila Lirio Marcelo, Nancy Lublin, Brian Lynch, Benjamin Mako Hill, Natalia Mehlman Petrzela, Michelle Michael, Geoff Mulligan, Nehkara Nikki Newhouse, Rainer Nõlvak, Alex Pentland, John Pinette, Shael Polakow-Suransky, Ai-jen Poo, Katie Radford, Thomas Reese, Jay

Rogers, Robin Sather, Nathan Schneider, Michael Silberman, James Slezak, Lara Stein, Courtnie Swearingen, Madelon van Tilburg, Eric Topol, Chris Wanstrath, David Weinberger, Paul Wicks, Rob Wijnberg, David Willey.

Most of all, we want to thank the many people we have never met who have already engaged with this thinking, improved it, and used it in their own work to make more people more powerful.

Henry wants to thank:

My colleagues and the board of directors at the 92nd Street Y for their support and encouragement throughout the process. In particular, Asha Curran not only offered invaluable feedback on the book as it developed, but our work together has informed (and improved) the thinking that ended up on these pages.

My mentors Bob Dilenschneider, Howard Gardner, Emily Rafferty, and Rob Reich all offered more valuable guidance than they know. Most of all I am indebted to the Chairman of the Board, Mark Angelson. The team at Stanford PACS offered a warm incubator for my ideas and a huge boost to my efforts.

My bookish friends Matthew Bishop, Mike Green, and Peter Sims offered whiskey, sympathy, and much wisdom. My BFF Jon Boden showed the way through his own creative ambition, and my brother Will always reminded me that you've got to believe it to see it. On a daily basis Callie and Josiah showed me the true limits of power, and Colleen showed me there are no limits to love.

Finally my parents, who gave me more gifts than I can ever count.

Jeremy wants to thank:

The friends and colleagues who have inspired me and been true collaborators over the past fifteen years, making pivotal contributions to my early work and adventures in movement building: Alessandra Orofino, Alnoor Ladha, Andy Kuper, Anna Nolan, Andre Banks, Andrea Woodhouse, Brett Solomon, Brian Berkopec, Cameron Hepburn, Christina Brower, Dan Barasch, David Madden, Douglas Atkin, Emma Bloomfield, Emmy Suzuki Harris, Erika Johansson, Evan Thornley, Galit Gun, Henry Donahue, James

Slezak, Lauren Rodman, Niko Canner, Sari Miller, Scott Heiferman, and Tim Dixon, among others.

My incredible partners at Purpose today—Josh Hendler, Dan Shannon, and Jessy Tolkan—and the whole team of Porps (with super-glue provided by Robbie Ross) who have supported me during the writing of this book, challenging my thinking and spurring me on.

My family, including my big sister Ingrid Bass and my brother Ralph Heimans and his wife, Tami, and especially my parents, Frank and Josette. My mother's lifelong love, support, and occasional reminders to work hard and stay focused on what matters, usually delivered in French or, if more urgent, Arabic. My father's deep commitment to justice and decency as a Holocaust survivor who took his experience and looked outward, telling the stories of other injustices as a documentary filmmaker and oral historian, and who gave me a model for independence and entrepreneurialism.

And finally Brock Forsblom. His kindness and connection made this book possible, and sustain me every day.

NOTES

Chapter 1: Welcome to the New Power World

1 "ability to produce intended effects": Bertrand Russell, *Power: A New Social Analysis* (London: Allen and Unwin, 1938).

2 In fact, according to *Quartz*: Ashley Rodriguez, "How Powerful Was Harvey Weinstein? Almost No One Has Been Thanked at the Oscars More," *Quartz*, October 13, 2017.

2 His films garnered over: Madeline Berg, "After Expulsion from the Academy, Here Are All of Harvey Weinstein's 81 Oscar Wins," *Forbes*, October 13, 2017.

2 The Queen made him: Alex Ritman, "British Government Under Pressure to Strip Harvey Weinstein of Honorary CBE," *Hollywood Reporter*, October 11, 2017.

2 He controlled the media: Jim Rutenberg, "Harvey Weinstein's Media Enablers," *New York Times*, October 6, 2017.

2 He hired: Ronan Farrow, "Harvey Weinstein's Army of Spies," *The New Yorker*, November 6, 2017.

3 Once there, as she reported to CNN: Emanuella Grinberg and Janet DiGiacomo, "Amid Harassment Allegations, a Sisterhood Forms to Take Down James Toback," CNN, October 30, 2017.

3 Together they formed: Ibid.

3 Within days of the article being published: Doha Madani, "Over 300 Women Chime In After L.A. Times Details Director's Sex Abuse Reputation," *Huffington Post*, October 27, 2017.

3 In just one day: "More Than 12M 'Me Too' Facebook Posts, Comments, Reactions in 24 Hours," CBS News, October 17, 2017.

3 In France it became: Eleanor Beardsley, "Instead of #MeToo, French Women

Say 'Out Your Pig,'" NPR, November 3, 2017; Sasha Lakach, "#MeToo Has Gone Global," Mashable, October 20, 2017.

4 The UK defense minister: Joshua Lowe, "#MeToo Sexual Assault and Harrassment Scandal in British Parliament Causes Minister to Resign," *Newsweek,* November 1, 2017.

4 The European Parliament: Milan Schreuer, "A #MeToo Moment for the European Parliament," *New York Times,* October 25, 2017.

4 India debated an effort: "#MeToo: 'Sexual predators' List Divides Indian Feminists," DW, November 8, 2017.

4 An article in *China Daily*: Nicholas Moore, "#MeToo: Fighting Workplace Harassment in China," CGTN, October 30, 2017.

4 #MeToo had been born: Justin Carissimo, "Creator of Original 'Me Too' Campaign Speaks Out," CBS News, October 17, 2017.

4 Everyone from enterprising: Heather Schwedel, "The Jewelry Designer Who's Already Selling #MeToo Necklaces Explains Herself," Slate, October 20, 2017.

5 This patient had learned: PatientsLikeMe, July 2017. www.patientslikeme .com.

5 "out of desperation": Letitia Browne-James, discussion with authors, June 28, 2017. Quotes and other information about her case are also from this source and at Tampa General Hospital Health News. www.tgh.org.

6 She was described: Atika Shubert and Bharati Naik, "CNN Exclusive: From Glasgow Girl to 'Bedroom Radical' and ISIS Bride," CNN, September 5, 2014.

6 "If I could advise you": Lauren Crooks, "'Bring good quality bras': Scots Jihadi Bride Writes Suitcase Checklist for Schoolgirls Wanting to Join Islamic State Fanatics," *Daily Record,* August 1, 2015.

7 When three normal and well-liked girls: Katrin Bennhold, "Jihad and Girl Power: How ISIS Lured 3 London Girls," *New York Times,* August 27, 2015.

7 It printed thousands: Luiz Martinez, "US Drops Anti-ISIS Leaflets over Syria," ABC News, March 26, 2015.

7 "Think Again Turn Away!": Rita Katz, "The State Department's Twitter War with ISIS Is Embarrassing," *Time,* September 16, 2014.

8 A popular thread: Reddit, "90's Kids, What's Something You Did When You Were in School That Youths of Today Wouldn't Understand?," Reddit, June 15, 2015. www.reddit.com.

9 New power models work more like Minecraft: Matt Peckham, "'Minecraft' Is Now the Second Best-Selling Game of All Time," *Time,* June 2, 2016.

12 He retweeted his most extreme: Taylor Wafford, "Donald Trump Retweets Racist Propaganda," *Newsweek,* November 23, 2015.

12 He offered to pay: Alan Rappeport, "Donald Trump Says He May Pay Legal Fees of Accused Attacker from Rally," *New York Times,* March 13, 2016.

12 Henry launched #GivingTuesday: #GivingTuesday, July 2017. www.giving tuesday.org.

12 Jeremy created a technology-powered: Purpose, July 2017. www.purpose .com.

Chapter 2: Thinking Old Power, Thinking New Power

14 "HQ are telling us": Hila Lifshitz-Assaf, "Dismantling Knowledge Boundaries at NASA: From Problem Solvers to Solution Seekers," *Administrative Science Quarterly,* 2017. https:doi.org/10.1177/0001839217747876.

14 The story of NASA's efforts: Ibid.

15 Three thousand people: Ibid.

15 And not only did the crowd: Ibid.

15 And now Bruce Cragin: Ibid.

15 "This puts you on the edge": Ibid.

15 "The tensions, debates and forces": Hila Lifshitz-Assaf, discussion with authors, August 24, 2016.

16 "fearing it might be transformed": Lifshitz-Assaf, "Dismantling Knowledge Boundaries."

16 "withholding the fact": Ibid.

16 "open NASA": Ibid.

16 "Space Apps hackathon": Space Apps Challenge, July 2017. www.2017.space appschallenge.org.

17 "is really intrinsic": Lifshitz-Assaf, "Dismantling Knowledge Boundaries."

17 "why they joined NASA": Hila Lifshitz-Assaf, discussion with authors.

17 "I didn't ask them about *them*": Ibid.

18 "The lab is my world": Lifshitz-Assaf, "Dismantling Knowledge Boundaries."

20 "We've heard people say": Peg Tyre, "Beyond School Supplies: How DonorsChoose Is Crowdsourcing Real Education Reform," *Fast Company,* February 10, 2014.

20 The site has seen over two million: DonorsChoose.org, July 2017. www .donorschoose.org.

20 "We have vested school boards": Tyre, "Beyond School Supplies."

21 "an opt-in society": Kevin Roose, "Silicon Valley's Secessionist Movement Is Growing," *New York Magazine,* October 21, 2013.

21 "Yelp for Drugs": Peter Kafka, "Balaji Srinivasan, Who May Run the FDA for Trump, Hates the FDA," *Recode,* January 14, 2017.

22 In a leaked document: Alyson Shontell, "A Leaked Internal Uber Presentation Shows What the Company Really Values in Its Employees," *Business Insider,* November 19, 2014.

22 "cooperative contexts proved": Geoffrey James, " 'Collaboration' Creates Mediocrity, Not Excellence, According to Science," *Inc.,* April 14, 2017.

22 "I mean, politics is like sausage": "HRC Paid Speeches," WikiLeaks, January 25, 2016.

23 "Noah has had both deep": Noah Dyer for Governor, "Scandal and Controversy," July 2017. www.noahdyer.com.

24 "It really comes down": "The Maker," Vimeo video, posted by "Patrick Kehoe," 4:43, January 18, 2016. www.vimeo.com.

24 "GynePunks": Doug Bierend, "Meet the GynePunks Pushing the Boundaries of DIY Gynecology," *Vice*, August 21, 2015; Wattpad, July 2017. www.wattpad.com.

24 In fact, over the last decade: Edelman, "2017 Edelman Trust Barometer Annual Global Study," 2017. www.edelman.com.

25 "people in this country": Michael Gove, "Britons 'Have Had Enough of Experts,'" *Sky News*, June 21, 2016.

25 "It's the road back": Chris York, "Professor Brian Cox Says Michael Gove's 'Anti-Expert' Stance Is the 'Road Back to the Cave,'" *Huffington Post*, February 7, 2016.

25 Robert Putnam's now-classic book: Robert Putnam, *Bowling Alone: The Collapse and Revival of American Community* (New York: Simon and Schuster, 2001).

26 "We're anti-excellence": Pastor Nadia Bolz-Weber, discussion with authors, October 30, 2015. All Bolz-Weber quotes in this section are from this source.

27 "Among the youngest Millennials":"'Nones' on the Rise: One-in-Five Adults Have No Religious Affiliation," *Pew Research Center*, 2012.

29 It has an amazingly effective: Airbnb, "About Us," July 2017. www.airbnb.com.

30 It has shown a strong commitment: "The Footprint Chronicles," July 2017. www.patagonia.com/footprint.

30 Impressively, in 2017 it claimed: Zainab Mahmood, "Guardian Creates Chief Customer Officer Role as It Claims 230,000 Paying Members," *Press Gazette* (UK), June 19, 2017.

30 It now pulls in: David Bond, "Guardian Relies on Readers' Support to Stave Off Crisis," *Financial Times*, May 13, 2017.

30 its database, "The Counted": Emma Howard and Damian Carrington, "Everything You Wanted to Ask About the *Guardian*'s Climate Change Campaign," *The Guardian*, March 16, 2015; "The Counted," *The Guardian*, July 2017. www.theguardian.com.

Chapter 3: From Sound Bites to Meme Drops: How Ideas Spread

33 "Who are you?": "Neck and Nominate," YouTube video, 1:22, posted by "Will Green," November 29, 2012, www.youtube.com.

33 In the video posted on YouTube: "24 HR cold water challenge," YouTube video, 2:34, posted by "Jessica Lagle," March 8, 2014, www.youtube.com.

33 In this iteration, firefighters: "Jennifer Begley's Coldwater Challenge Lexington Fire Department for Joe," YouTube video, 1:32, posted by "Jennifer Begley," May 31, 2014, www.youtube.com.

34 A piece on *KYForward*: Lexington Fire Department, "Firefighters, News Anchors, Mayor Accept Cold Water Challenge Supporting Fellow Fireman," *KYForward*, June 3, 2014.

34 This moment has earned: "CK Ice Bucket Challenge," YouTube video, 0:25, posted by "Chris Kennedy," July 15, 2014, www.youtube.com.

34 But the true catalytic event: Pete Frates's Facebook page, July 31, 2014, www.facebook.com.

34 "It was very chilly": Alex Finnis, "Pensioner, 102, Celebrates Becoming Britain's Oldest Ice 'Bucketeer' with a Glass of Whisky," *Daily Mail*, August 26, 2014.

35 The ALS Association ultimately raised: ALS Association, *Annual Report 2014: Finding Connections, Finding a Cure,* 2014. www.alsa.org.

35 "183% increase": Ibid.

37 In their terrific best-selling book: Chip and Dan Heath, *Made to Stick: Why Some Ideas Survive and Others Die* (New York: Random House, 2007).

37 What these examples share: Ibid., 16–18.

39 Manju Latha Kalanidhi: Arpita Aneja, "Rice, Not Ice: India's Answer to the Ice Bucket Challenge," *Time,* August 26, 2014.

39 Patrick Stewart of *Star Trek*: "Pstew's Ice Bucket Challenge," YouTube video, 0:56, posted by "Sunny Ozell," August 22, 2014, www.youtube.com.

40 "What City Should You *Actually* Live In?": Ashly Perez, "What City Should You Actually Live In?," *BuzzFeed*, January 16, 2014.

40 "whether a certain dress": Cates Holderness, "What Colors Are This Dress?," *BuzzFeed*, February 25, 2015.

40 "17 Times Double-Jointed People": Caroline Kee, "17 Times Double-Jointed People Took It Way Too Far," *BuzzFeed*, August 22, 2015; "100 Most Australian Words": Chris Rodley, "100 Most Australian Words of All Time," *BuzzFeed*, June 15, 2015.

40 The hungry may be part: BuzzFeed Tasty Facebook page, July 2017, www.facebook.com.

40 "envy of the media world": Chris Matthews, "Here's Why Buzzfeed Could Be Worth $1.5 Billion," *Fortune*, July 31, 2015; "The Most Innovative Companies of 2017," *Fast Company,* February 2017; Noah Robinschon, "How BuzzFeed's Jonah Peretti Is Building a 100-Year Media Company," *Fast Company,* February 16, 2016.

40 "If your goal": Ben Smith, "Why BuzzFeed Doesn't Do Clickbait," *BuzzFeed*, November 6, 2014.

41 "personal dialect map": "How Y'all, Youse and You Guys Talk," *New York Times,* December 21, 2013.

41 "We want to create value": Abhinn Shreshtha, "You Cannot Charge for Vanilla Content," *Exchange 4 Media,* March 8, 2017.

41 "If None of These Pictures": Dave Stopera, "If None of These Pictures Make You Say 'What the Fuck,' Nothing Will," *BuzzFeed*, May 7, 2017.

41 "strange and radical": "Let Them Wed," *The Economist,* January 4, 1996.

42 It divided those who received: Inga Kiderra, "Facebook Boosts Voter Turnout," UC San Diego News Center, September 12, 2012.

42 They published their study: Zoe Corbyn, "Facebook Experiment Boosts US Voter Turnout," *Nature,* September 12, 2012. Quotes and data on the experiment that follow also come from this source.

43 "People actually begin changing": Alex (Sandy) Pentland, discussion with authors, August 14, 2015.

43 "Our data show": Jack Krawczyk and Jon Steinberg, "How Content Is Really Shared: Close Friends, Not 'Influencers,'" *Advertising Age,* March 7, 2012.

44 *Humans of New York*: Humans of New York, July 2017. www.humansofnew york.com.

44 "We're talking about how important": *Humans of New York*'s Facebook page, April 26, 2017, and May 1, 2017, www.facebook.com/humansofnewyork.

44 Stanton has 17 million Facebook followers: Ibid., July 2017.

45 The 2015 event saw PayPal: Sophia Rocher, "Paypal Smashes World Record for Most Money Raised Online for Charity in 24 Hours with $45.8M Campaign," *Guinness World Records News,* January 13, 2016. www.guinnessworld records.com

45 "Arguably, it's become": Micah Sifry, "The Most Successful Civic Tech Culture Hack of the Decade?," *Medium,* December 4, 2015.

45 It set an ambitious goal: University of Michigan, "Giving Blueday Impact," December 2016. www.givingblueday.org.

45 The organization Dress for Success: Dress for Success, "Dress for Success Turns #GivingTuesday Into #GivingShoesDay," October 30, 2015. www .dressforsuccess.org.

46 There is a #GivingZooDay: GivingZooDay, July 2017. www.givingzooday .org.

46 Together they raised over $5 million: Madeline Turner, "How Baltimore Became the Most Generous City in America on Giving Tuesday," *npEN-GAGE* (blog), November 4, 2015. www.npengage.com.

46 "Doa Sorocaba": Doa Sorocaba, July 2017. www.doasorocaba.com.br.

46 In South America, #GivingTuesday: see UnDiaParaDarAR, July 2017. www .undiaparadar.org.ar.

46 Singapore has changed #GivingTuesday: GivingWeek, July 2017. www .givingweek.sg.

46 In Russia it is #ЩедрыйВторник, July 2017. www.givingtuesday.ru.

46 It won the title: Lauren Johnson, "Taco Bell's Cinco de Mayo Snapchat Lens Was Viewed 224 Million Times," *Adweek,* May 11, 2016.

47 *Adweek* reported that: Ibid.

47 "The decision to end": Muscular Dystrophy Association, "MDA Telethon Ends Historic Run, Urgent Fight for Families Continues," May 1, 2015. www .mda.org.

48 Fundraising for the ALS Association: Amy Nordrum, "Ice Bucket Challenge

2015: Can the ALS Association Turn Last Year's Viral Phenomenon into an Annual Fundraiser?," *International Business Times*, August 26, 2015.

48 "Sometimes, it's better to just leave": Khaleda Rahman and Victoria Allen, "The Private School Jihadist," *Daily Mail*, September 2, 2014.

49 "ISIS poster girl": James Cook, "Glasgow 'Jihadist' Aqsa Mahmood Denies Recruiting London Girls," *BBC News*, March 16, 2016.

49 "jihadi brides": Lizzie Dearden, "Isis 'Jihadi Brides' Trying to Radicalise Girls and Encourage UK Terror Attacks Online as They Remain Trapped in Syria," *The Independent*, August 13, 2016.

49 "Follow the example of your Brothers": National Post Staff, "Inside the Life of a 20-Year-Old Scottish Woman Who Ran Away to Become a Hardline Supporter of ISIS," *National Post*, September 4, 2014.

49 "to send secret advice": Lauren Crooks, "Scots Jihadi Bride Uses Secret Online Message Service to Recruit Fighters for Islamic State," *Daily Record*, February 8, 2015.

49 "The first phone call": Ashley Fantz and Atika Shubert, "From Scottish Teen to ISIS Bride and Recruiter: The Aqsa Mahmood Story," CNN, February 24, 2015.

50 It has invested in old power: Rukmini Callimachi, "A News Agency with Scoops Directly from ISIS, and a Veneer of Objectivity," *New York Times*, January 14, 2016.

50 "#YODO—You only die once": Colin Lecher, "ISIS Is Waging a Terrifying Web Propaganda War," *The Verge*, June 23, 2014.

50 *The Guardian* reported in 2014: Patrick Kingsley, "Who Is Behind Isis's Terrifying Online Propaganda Operation?," *The Guardian*, June 23, 2014.

50 "The picture that said": Ibid.

50 "Now, in the age of YouTube": Steve Rose, "The Isis Propaganda War: A Hi-Tech Media Jihad," *The Guardian*, October 7, 2014.

51 "Think Again Turn Away!" campaign: Katz, "State Department's Twitter War."

51 But the reality: Adam Edelman, "State Department's 'Embarrassing' 'Think Again Turn Away' Twitter Campaign Could Actually Legitimize Terrorists: Expert," *NY Daily News*, September 14, 2016.

51 The jihadis often: Katz, "State Department's Twitter War."

51 "Welcome to the 'Islamic State' Land (ISIS/ISIL)": Ibid.

51 "You need to find a way": U.S. Government Committee on Oversight and Government Reform, "Radicalization: Social Media and the Rise of Terrorism: Hearing Before the Subcommittee on National Security of the Committee on Oversight and Government Reform," U.S. Government Publishing Office, October 28, 2015, 11.

51 "network of positive messengers": Ibid.

52 "push back on online hate": EdVenture Partners, "Peer to Peer," July 2017. www.edventurepartners.com.

52 It mobilized the masses: U.S. State Department's Bureau of Educational

and Cultural Affairs, "FATE: From Apathy to Empathy—P2P #Challenge Extremism," March 4, 2016. www.eca.state.gov.

52 Another came from a group of Finnish students: Embassy of Finland, "Finnish Team Wins U.S. Competition on Fighting Extremism," February 10, 2016. www.finland.org.

52 A U.S. team created: "Snapchat: UNGA LGBT Core Group Event and P2P Extremism Challenge," YouTube video, 3:33, posted by "U.S. Department of State," September 22, 2016, www.youtube.com.

52 A class from Azerbaijan: EdVenture Partners, "Peer to Peer Challenging Extremism—Spring 2016 Awards," July 27, 2016. www.edventurepartners .com.

52 As one promising response: "The Redirect Method," July 2017. www .redirectmethod.com.

Chapter 4: How to Build a Crowd

55 This ad for Pepsi: Roisin O'Connor, "Kendall Jenner and Pepsi Face Backlash over Advert Showing Model Giving Can to Police Officer Mid-Protest," *The Independent*, April 5, 2017.

55 "Pepsi's Kendall Jenner Ad": Madison Malone Kircher, "Pepsi's Kendall Jenner Ad Sparks Mass Mobilization of Mean, Funny Tweets," *New York* magazine, April 5, 2017.

56 So in August of 2005: GetUp! Action for Australia, July 2017. www.getup .org.au.

56 They made a TV ad: "Get Up!," YouTube video, 0:46, posted by "GetUp! Australia," August 1, 2011, www.youtube.com.

56 "I'm not going to give up": Ibid.

57 Within days, tens of thousands: GetUp! Australia, "GetUp Achievements," July 2017. www.getup.org.au.

57 "There are hundreds of emails": Mark Bannerman, "Website Hopes to Spark Political Interest," Australian Broadcasting Corporation, August 4, 2005.

57 When the votes for Parliament were counted: Paul Bibby, "Finally, Howard Admits McKew Has It," *Sydney Morning Herald*, December 12, 2007.

58 In the years since it was launched: GetUp! Australia, "GetUp Achievements."

58 It has raised tens of millions: GetUp! Australia, "Powering GetUp," July 2017. www.getup.org.au.

58 "probably [Australia's] most important": Kathy Marks, "Exclamation Politics: GetUp!," *The Monthly*, October 2010.

58 "is to politics what Uber": Nick Cater, "Incoherent Get Up! Just Goes Along for a Disruptive Ride," *The Australian*, May 23, 2017.

60 The online crafting marketplace: "About Etsy," July 2017. www.etsy.com.

60 One of them, Rob Kalin: Max Chafkin, "Can Rob Kalin Scale Etsy?," *Inc.*, April 2011.

60 blogging platforms like LiveJournal: LiveJournal, July 2017. www.live journal.com.

60 Some of the most passionate members: Morgan Brown, "Etsy: [Ideas Inside] The Story of Etsy's Crafty Growth to IPO and a $2 Billion Valuation," *Growth-Hackers*, 2015.

60 "Etsy's core mission": Ibid.

62 The "About Us" section: GetUp!, "About Us," July 2017. www.getup.org.au.

63 "creating a world where": Douglas Atkin, email to the authors, November 3, 2017.

63 "This Tumblr shows": Joe Berkowitz, "This Tumblr Shows Everything Airbnb's New Logo Looks Like in Addition to a Vagina," *Fast Company*, July 21, 2014.

63 "Most brands would send": Austin Carr, "Airbnb Unveils a Major Rebranding Effort That Paves the Way for Sharing More Than Homes," *Fast Company*, July 16, 2014.

63 "optimal distinctiveness": M. B. Brewer, "The Social Self: On Being the Same and Different at the Same Time," *Personality and Social Psychology Bulletin* 17 (1991): 475–482.

63 "We used to take belonging for granted": Brian Chesky, "Belong Anywhere," Airbnb blog, July 2017. www.airbnb.com.

64 "Take me to the Hilton": "About," July 2017. www.hilton.com.

65 "greatest achievements": Chuck Gates, interview with authors, June 2017.

66 It repositioned the 144-year-old institution: National Guild, "92nd Street Y Named a 2016 Most Innovative Company," *National Guild*, February 22, 2016; "92nd Street Y Receives $15 Million for Innovation and Social Impact," *Philanthropy News Digest,* May 27, 2016.

67 The "dating" app Tinder: Tinder, July 2017. www.tinder.com.

68 "Jan Lokpal": Shanti Bhushan, "Jan Lokpal Bill and Parliament," *The Hindu,* September 6, 2011.

68 In early April 2011: Jason Burke, "Indian Activist Anna Hazare Refuses to End Hunger Strike," *The Guardian,* April 7, 2011.

69 So Hazare got himself: Jonathan Arp, "Call to Action," *Makeshift,* Fall 2013.

69 When Hazare provided: /The Rules, "Changing the World with Millions of Missed Calls," *Social Tech Guide,* May 9, 2014.

70 "weak ties": Malcolm Gladwell, "Small Change," *New Yorker*, October 4, 2010.

70 Hazare's then campaign contacted: Burke, "Indian Activist Anna Hazare."

70 /The Rules, a group: /The Rules, "Changing the World."

71 TED, the conference ommunity: TED, July 2017. www.ted.com.

72 There have now been over 20,000: TED, "Organize a Local TEDx Event," July 2017. www.ted.com.

72 Perhaps the most inane example: Yo, July 2017. www.justyo.com.

72 By mid-June the app was: Alyson Shontell, "What's Happened to ~$7 Mil-

lion App 'Yo' Now That the Hype Has Died," *Business Insider,* September 26, 2014.

73 "stupidest, most addictive app": Alyson Shontell, "The Inside Story of Yo," *Business Insider,* June 21, 2014.

73 By September, just a few months after launch: Shontell, "What's Happened to ~$7 Million App."

73 "civilization is a drama": John W. Gardner, "The Tasks of Leadership," *Project Kaleidoscope* 4 (2005), 2.

74 The tool was a product of Meu Rio: Joshua Goldstein, "Meu Rio," *Civic Media Project,* July 2017. www.civicmediaproject.org.

74 More than one in ten: Rachel Glickhouse, "Interview with Meu Rio's Alessandra Orofino," *Rio Gringa* (blog), May 1, 2013. www.riogringa.typepad .com.

75 As soon as the bulldozers: Liza Booth, "A Brazilian App Gives Power to the People," *ZDNet,* July 7, 2014.

75 And after six more months of persistent: Glickhouse, "Interview with Meu Rio's Alessandra Orofino."

76 If GetUp won: GetUp! Australia, "Children Out of Detention," July 2017. www.getup.org.au.

76 In 2010, Digg announced "version 4": Charles Arthur, "Digg Loses a Third of Its Visitors in a Month: Is It Deadd?," *The Guardian,* June 3, 2010.

76 "quit Digg day": Craig Kanalley, "Angry Users SLAM Digg with Links from Rival Reddit," *Huffington Post,* August 30, 2010.

77 "giving power back to the people": Alexis Ohanian, "An Open Letter to Kevin Rose," *Making the World Suck Less* (blog), 2010. www.alexisohanian .com.

77 The great Digg exodus: Charles Arthur, "Digg Sold for Just $500,000," *The Guardian,* July 13, 2012.

77 "you can still dunk": Oreo Cookie, Twitter post, February 3, 2013, 8:48 p.m., www.twitter.com/oreo.

77 "won the Super Bowl": Angela Watercutter, "How the Oreo Won the Marketing Super Bowl with a Timely Blackout Ad on Twitter," *Wired,* February 2, 2013.

78 They returned the funds: "The Girl Scouts' #ForEVERYGirl campaign," *Washington Post,* June 30, 2015.

78 The scouts raised $338,000: Henry Gass, "For Transgender 9-Year-Old, a Very Girl Scout Lesson," *Christian Science Monitor,* February 3, 2016.

78 "I participated in 15 years of Girl Scouting": #ForEVERYGirl's Indiegogo page, "Girl Scouts Is #ForEVERYGirl," July 2017. www.indiegogo.com.

79 Research on crowdfunding: Amy Yeh, "New Research Study: 7 Stats from 100,000 Crowdfunding Campaigns," *Indiegogo* (blog), October 6, 2015. www .go.indiegogo.com.

79 It sees that loss: Kaja Whitehouse, "Attorney General's NYC Battle with Airbnb Targets Widowed Grandma," *New York Post,* November 13, 2013.

79 We are witnessing the rise of "WeWashing": Lee-Sean Huang, "#WeWashing: When 'Sharing' Is Renting and 'Community' Is a Commodity," *Huffington Post*, May 16, 2015.

80 In one video, the parents: "DaddyOFive Invisible Ink Prank," YouTube video, 6:25, posted by "Polly Anner," April 18, 2017, www.youtube.com.

80 "Once people started watching us": Clint Davis, "YouTube Star DaddyO-Five Loses Custody of Kids After Complaints over 'Prank' Videos," *NBC26*, May 2, 2017.

Chapter 5: What Makes New Power Communities Work (and Why They Sometimes Don't)

82 "Dear Reddit": qgyh2, "Dear Reddit, you are starting to suck.," Reddit, July 3, 2015. www.reddit.com.

83 "front page of the internet": Arjun Kharpal, "Reddit: The 'Front Page of the Internet' Wants to Be a Billion-Dollar Business," CNBC, June 16, 2016.

83 More than 21 million votes: Michelle Brous, "Reddit: Managing the 'Front Page of the Internet,'" *Harvard Business School Digital Innovation and Transformation*, March 20, 2017. www.digit.hbs.org.

83 Reddit also has a big impact on politics: Trevor Martin, "Dissecting Trump's Most Rabid Following Online," *FiveThirtyEight* (blog), March 23, 2017. www .fivethirtyeight.com.

83 "morbidly beautiful things": Reddit, "New Subreddits by Date," July 2017. www.redditmetrics.com.

83 The "Reddit rules": Reddit, "Rediquette," July 2017. www.reddit.com.

84 She was beloved: Hayley Tsukayama, "Who Is Victoria Taylor, the Woman at the Heart of the Reddit Revolt?," *Washington Post*, July 6, 2015.

84 "When my wife got pregnant": Brian Lynch, discussion with authors, November 19, 2015.

84 On July 2, 2015: Sarah Burh, "Reddit: 'We Screwed Up,'" *TechCrunch*, July 6, 2015.

85 Other mods said, "'We'll go black'": Brian Lynch, discussion with authors.

85 "Then after that it just": Ibid.

85 A rolling blackout: Margaret Hartmann, "What Really Caused the Reddit Revolt?," *New York* magazine, July 15, 2015.

85 "We screwed up": Burh, "Reddit: 'We Screwed Up.'"

86 "Reddit's new age of censorship": Billy Johnson, "Step Down as CEO of Reddit Inc.," Change.org, July 2017. www.change.org.

86 "We will ban subreddits": Reddit, "Removing Harassing Subreddits," June 10, 2015. www.reddit.com.

86 "toxic 0.2%": Jessi Hempel, "Inside Reddit's Plan to Recover from Its Epic Meltdown," *Wired*, October 6, 2015.

87 "a manipulative individual": Johnson, "Step Down as CEO of Reddit Inc."

90 "You got to build a community": Brian Lynch, discussion with authors.

90 "Reddit is my community center": Courtnie Swearingen, discussion with authors, November 20, 2015.

90 The tension between: Wimali_Stebox, "What's Vote Brigading, and Why Is It Illegal," Reddit, April 30, 2014. www.reddit.com.

91 "everyone's private driver": Felix Salmon, "The Economics of 'Everyone's Private Driver,'" *Medium*, June 1, 2014.

91 "your friend with a car": Dimosthenis Kefallonitis, "Lyft.me, Your Friend with a Car," *Consumer Value Creation*, January 22, 2014.

91 Uber is defined by its remoteness: Alanna Petroff, "The Rise and Fall of Uber CEO Travis Kalanick," *CNNMoney*, June 21, 2017.

91 "The reason Uber": Chris Smith, "Uber Wants to 'Get Rid of the Dude in the Car' with Driverless Taxi Service," *TechRadar*, May 8, 2014.

92 It all began when Uber: Caroline O'Donovan, "Uber Just Cut Fares in 80 North American Cities," *BuzzFeed*, January 9, 2016.

92 "In true Uber fashion": Harry Campbell, "Uber to Cut Rates in More Than 100 Cities," *The Rideshare Guy* (blog), January 8, 2016. www.therideshareguy.com.

92 "Even with [our] better service": John Zimmer, "Standing Together: Community Update from John," *The Hub* (blog), February 2, 2016. www.thehub .lyft.com

93 "the bonds you create": Ibid.

93 "I love @Lyft drivers": Cori Online, "#ThankYourLyftDriver," *Lyft* (blog), January 31, 2016. www.blog.lyft.com.

94 "As a new Lyft driver": Harry Campbell, discussion with authors, February 16, 2016.

94 "basically get your entire commission": Timothy B. Lee, "Lyft Says Its Drivers Can Make $35 an Hour," *Vox*, December 17, 2014.

94 "One of the things that Uber drivers": Harry Campbell, discussion with authors.

95 "shared norms, values": Arun Sundararajan, "What Airbnb Gets About Culture That Uber Doesn't," *Harvard Business Review*, November 27, 2014. Other quotes in this paragraph are from the same source.

95 In a 2017 survey: Laura Sydell, "Survey Finds Lyft Drivers Happier Than Uber, Though Pay Has Declined," NPR, January 21, 2017.

96 To take another example, YouTube's Partner Program: Todd Spangler, "YouTube Standardizes Ad-Revenue Split for All Partners, but Offers Upside Potential," *Variety*, November 1, 2013.

96 And Etsy's long-standing commitment: Nicole Leinbach-Reyhle, "Etsy vs. Amazon: Different Ideas of What Is 'Handmade,'" *Forbes*, October 10, 2015.

97 Their network allows anyone: Meetup, July 2017. www.meetup.com.

97 So now Meetup's primary business model: Meetup, "Organizer Subscription Pricing," July 2017. www.meetup.com.

97 "Kickstarter is not a store": Yancey Strickler, Perry Chen, and Charles Adler, "Kickstarter Is Not a Store," *The Kickstarter* (blog), September 20, 2012. www .kickstarter.com.

99   In fact, the celebrated venture capital firm: Alyson Shontell, "Why Legendary Investor Fred Wilson Didn't Invest in Airbnb When It Was Just a Tiny Startup," *Business Insider*, March 21, 2014.

99   "When we rate each other": Tom Slee, "The Shape of Airbnb's Business (II)," *Tom Slee* (blog), June 9, 2014. www.tomslee.net.

100   "nonblack hosts are able": Ben Edelman and Michael Luca, "Digital Discrimination: The Case of Airbnb.com," *Harvard Business School Publishing*, January 10, 2014, 2.

100   Further research by Edelman, Luca, and Dan Svirsky: Ben Edelman, Michael Luca, and Dan Svirsky, "Racial Discrimination in the Sharing Economy: Evidence from a Field Experiment," *Harvard Business School Publishing*, September 4, 2016.

100   "profile photos are essential": Sam Levin, "Airbnb Adopts New Rules in Effort to Fight Racial Discrimination by Hosts," *The Guardian*, September 8, 2016.

100   Initially created as a workaround: Chris Garces, "People's Mic and Democratic Charisma: Occupy Wall Street's Frontier Assemblies," *Focaal: Journal of Global and Historical Anthropology* 2013, no. 66 (Summer 2013): 88–102.

101   "The human megaphone": Craig Calhoun, "Occupy Wall Street in Perspective," *British Journal of Sociology* 64, no. 1 (2013): 30.

101   "Three weeks into the occupation": Andrew Cornell, "Occupy Wall Street and Consensus Decision Making: Historicizing the Preoccupation with Process," *Is This What Democracy Looks Like?* (blog), October 8, 2013. www .what-democracy-looks-like.org.

102   "That was our impetus": Ben Keesey, discussion with authors, September 24, 2015.

102   "in search of our story": Steve Bramucci, "Why the 'Kony 2012' Campaign Should Make Travelers Uncomfortable," *Huffington Post*, March 13, 2012.

103   In total, they launched 16 tours: Invisible Children, "National Tour," July 2017. www.invisiblechildren.com.

103   Called "Roadies": Ibid.

103   "you're part of something huge": "Be a Roadie," Vimeo video, 3:20, posted by "Invisible Children," October 13, 2009. www.vimeo.com.

103   "Roadie teams could see": Keesey, discussion with authors.

104   It cultivated the network: Jessica McKenzie, "Inside Invisible Children's Massive Grassroots Network," *Mobilization Lab*, January 15, 2015.

104   "In order for your city": Zach Barrows, discussion with authors, October 6, 2015.

104   This led to 80,000 kids: McKenzie, "Inside Invisible Children's Massive Grassroots Network."

104   Over 90 percent had talked: Beth Karlin, "Power Through Participation: Impacts of Youth Involvement in Invisible Children," presented at the 53rd Annual International Studies Association Convention, April 1, 2012, 11.

104   "It was like": Barrows, discussion with authors.

Chapter 6: What Makes New Power Communities Work
(and How They Fuse with the Wider World)

106   They called this film: *Kony 2012*, YouTube video, 29:59, posted by "Invisible Children," March 5, 2012, www.youtube.com.

107   "the most viral video in history": Samantha Grossman, "'Kony 2012' Documentary Becomes Most Viral Video in History," *Time*, March 12, 2012.

107   Surveys suggested that more: Lee Raine, Paul Hitlin, Mark Jurkowitz, Michael Dimock, and Shawn Neidorf, "The Viral Kony 2012 Video," Pew Research Center, March 15, 2012.

107   "20 culture makers and 12 policymakers": *Kony 2012*.

107   "make Kony famous": Christie D'Zurilla, "Kony 2012: Which 20 Celebrities Are Targeted to Help?," *Los Angeles Times*, March 8, 2012.

107   "This movement did not emerge": David Campbell, "Kony 2012: Networks, Activism and Community," *David Campbell* (blog), March 16, 2012. www.david-campbell.org.

107   "Thanks tweeps for sending": D'Zurilla, "Kony 2012."

107   None other than Bono: Malaka Gharib, "Bono Comments on Invisible Children's Kony 2012 Campaign," *One*, March 12, 2012. www.one.org.

107   So did Jason Russell: Erin Carlson, "'Kony 2012' Creator Jason Russell Addresses Nude Breakdown on 'Oprah's Next Chapter' (Video)," *Hollywood Reporter*, October 8, 2012.

108   There was such volume: Polly Curtis and Tom McCarthy, "Kony 2012: What's the Real Story?," *The Guardian*, March 8, 2012.

108   A university student started: Visible Children's Tumblr page, July 2017. www.visiblechildren.tumblr.com.

108   "Cover the Night": Joshua Trujillo, "Kony 2012's 'Cover the Night' Continues Despite Controversy," *Seattle PI*, April 23, 2012.

108   Twenty-one thousand had signed up: Mike Hager, "Kony 2012 Campaign Fails to Go Offline in Vancouver," *Vancouver Sun*, April 21, 2012.

108   The Brisbane event: Chris Paine, "KONY 2012's Struggle to Remain Visible," *NewsComAU*, April 21, 2012.

109   "core model of hosting": Ben Keesey, "Working Myself Out of a Job: Lessons from Leading Invisible Children," *The Guardian*, December 17, 2015.

110   "When *Kony 2012* hit": Zach Barrows, discussion with authors, October 6, 2015.

110   For another example of this dynamic: Associated Press, "Livestrong Revenue Dropped Again in 2015, Trending Up Now," *USA Today*, April 26, 2017.

111   "We're in a political campaign": Sarah Kessler, "Taxi and Uber Drivers, Once Mortal Enemies, Join Forces in New Labor Dispute," *Fast Company*, February 19, 2016.

111   "When I first started driving": Ibid.

111 In 2016 it was fighting: Tess Townsend, "Uber Has Been Sued 46 Times in the Past Six Months," *Inc.*, July 6, 2016.

111 These tensions were behind: Elena Cresci, "#DeleteUber: How Social Media Turned on Uber," *The Guardian,* January 30, 2017.

111 Not only was Kalanick: Mike Isaac, "UBER C.E.O to leave Trump Advisory Council After Criticism," *New York Times*, February 2, 2017.

111 Weeks later, another surge: Susan J. Fowler, "Reflecting on One Very, Very Strange Year at Uber," *Susan J. Fowler* (blog), February 19, 2017. www.susan jfowler.com.

112 "deeply hurting": Tessa Berenson, "Uber to Users Deleting App Over Sexual Harassment Charges: We're 'Deeply Hurting,'" *Fortune*, February 24, 2017.

112 Kara Swisher and: Kara Swisher and Johana Bhuiyan, "A Top Uber Executive, Who Obtained the Medical Records of a Customer Who Was a Rape Victim, Has Been Fired," *Recode*, June 7, 2017.

112 Media, regulators, competitors: Mike Isaac, "For Uber's Travis Kalanick, Wrong Turns Led to a Forced Exit," *Economic Times*, June 22, 2017.

112 In 2016, Uber lost: Seth Fiegerman, "Uber Lost $2.8 Billion Last Year," CNN, April 14, 2017.

113 She was even featured: Ligia Friedman, discussion with authors, December 2, 2016.

113 "built the factories": Steve Adler, discussion with authors, February 4, 2017.

113 "The app is essentially a clone": Bryan Clark, "SXSW Showed Us the Future of Ride Sharing, and It's Not Uber," *TheNextWeb*, March 15, 2017.

113 "Built by Austin": Ride Austin, July 2017. www.rideaustin.com.

113 One hundred percent of the normal fee: Ride Austin, "Round Up," July 2017. www.rideaustin.com.

113 "Uber keeps its own power": Joe Deshotel, discussion with authors, November 14, 2016.

114 "Look, you like the model": Ibid.

114 By spring 2017, it was closing in on: Dan Solomon, "Hailing a Different Ride in Austin," *MIT Technology Review,* March 7, 2017; Ride Austin staff, "Ride Austin numbers," email to the authors, 2017.

115 Yet, as Deshotel reminded us: Deshotel, discussion with authors.

116 "Personally, I think the idea": Casey Newton, "Zuckerberg: The Idea That Fake News on Facebook Influenced the Election Is 'Crazy,'" *The Verge*, November 10, 2016.

116 Five months later Facebook reversed: Jen Weedon, William Nuland, and Alex Stamos, "Information Operations and Facebook," Facebook, April 27, 2017.

116 What is the psychological impact: Holly B. Shakya and Nicholas A. Christakis, "A New, More Rigorous Study Confirms: The More You Use Facebook, the Worse You Feel," *Harvard Business Review,* April 10, 2017.

Chapter 7: The Participation Premium

117 "traditional terrestrial options": Chris Roberts, "Letter from the Chairman," Roberts Space Industries, November 27, 2014. www.robertsspaceindustries .com.

118 The creator of the classic: Wolff Bachner, "Chris Roberts Returns to Game Design: Unveils 'Star Citizen' at GDC," *Inquisitr*, October 11, 2012.

118 "a living, breathing science fiction": "About the Game," July 2017. www .robertsspaceindustries.com.

118 In an hour-long presentation: Cloud Imperium, "Legendary Designer Chris Roberts Making Re-entry into PC Gaming Stratosphere with Star Citizen from Cloud Imperium," *Business Wire*, October 10, 2012.

119 "As someone who has played every Wing Commander": TwitchingCheese, "Roberts Space Industries GDC Panel Live—Chris Roberts of Wing Commander Fame Reveals His New Game," Reddit, 2013. www.reddit.com.

119 "a limited number of people": "Star Citizen Game Panel at GDC Austin October 10th 2012," YouTube video, 58:32, posted by "Star Citizen S," June 14, 2013, www.youtube.com.

119 "If this is something you're excited by": Ibid.

119 "Apparently our site": Ibid.

119 In the first twenty-four hours: "An Update from Chris Roberts," Star Citizen's Kickstarter page, October 19, 2012. www.kickstarter.com.

119 "Remember that the more we raise": Roberts Space Industries Star Citizen, "Comm-Link: An Update from Chris Roberts, 2012-10-19," October 19, 2012. www.starcitizen.tools.

120 Sixty dollars would get you: Star Citizen, "Backer Rewards," *Star Citizen Wiki*, July 2017. www.starcitizen.gamepedia.com.

120 "Spectrum Dispatch": "Spectrum Dispatch," July 2017. www.robertsspace industries.com.

120 "the stories and the creativity": Chris Roberts, "A Note from Chris Roberts," *Steam*, February 2011. www.steamcommunity.com.

121 "The battle is over": Star Citizen's Kickstarter page, "Thank You!—End of Campaign Transition," November 19, 2012. www.kickstarter.com.

121 In June, RSI celebrated: Brian Gaar, "Crowdfunding Raises $10 Million for Austin-Developed Game," *My Statesman*, June 11, 2013.

121 "CitizenCon": "Star Citizen CitizenCon 2014 Complete Show," YouTube video, 2:01:36, posted by "Star Citizen," October 14, 2014, www.youtube.com.

121 It generated $2 million: Eddie Makuch, "Star Citizen Funding Passes $23 Million," *GameSpot*, October 18, 2013.

121 "increase immersiveness": Phil Savage, "Star Citizen's Crowdfunding Total Shoots Past $50 Million," *PC Gamer*, August 18, 2014.

121 At $64 million: Preetam, "Star Citizen Smashes Past $64 Million Target, Introduces Space Pets," *NextPowerUp*, November 28, 2014.

121 "Please raise your hand": "Sandi's Opening Speech," YouTube video, 5:55, posted by "Mitauchi," October 10, 2015, www.youtube.com.

121 By December 2015: JD Alois, "Incredible: Star Citizen Tops $100 Million," *Crowdfunding Insider*, December 13, 2015.

123 "Several years from now": Chris Roberts, "Letter from the Chairman," Roberts Space Industries, July 20, 2015. www.robertsspaceindustries.com.

123 "I have already got my money's worth": Charlie Hall, "Star Citizen's FPS Module Delayed Indefinitely," *Polygon*, June 30, 2015.

124 In China, the smartphone vendor Xiaomi: Catherine Shu, "Xiaomi Confirms It Raised $1.1B At $45B Valuation," *TechCrunch*, December 29, 2014.

124 In 2015, powered by: Russell Flannery, "Xiaomi Breaks World Record for Online Mobile Phone Sales in a Day," *Forbes*, April 9, 2015.

125 Even its competitors have referred: Paul Mozur and Shanshan Wang, "The Rise of a New Smartphone Giant: China's Xiaomi," *New York Times*, December 14, 2014.

125 Throughout the year: Kaylene Hong, "Xiaomi's Social Media Strategy Drives Fan Loyalty, Books It $242m in Sales in 12 Hours," *TheNextWeb*, April 9, 2014.

125 An early 2012 interview: Hamish McKenzie, "Already Valued at $4B, China's Xiaomi Has Declared War on Apple," *Pando*, July 17, 2012.

125 And its fans weren't just voting: Ibid.

126 "I really enjoy the sense": Mozur and Wang, "The Rise of a New Smartphone Giant."

126 By 2015, around the time: Marcus Wolsen, "You May Never Use Xiaomi's Phones, but They'll Change Your Life Anyway," *Wired*, February 12, 2015.

126 Wanqiang Li: Kristie Wong, "Xiaomi and the Power of the Fan Economy," *Freshtrax*, September 22, 2014.

126 "having a research and development team": Ibid.

126 Company lore has developed: Jessica Dolcourt, "Xiaomi's Key to Smartphone Success: 'Be Friends with Our Fans,'" *CNET*, February 13, 2015.

126 "IKEA effect": Michael I. Norton, Daniel Mochon, and Dan Ariely, "The 'IKEA Effect': When Labor Leads to Love," *Harvard Business School Publishing*, 2011.

126 "Builders valued their": Ibid., 11.

127 His fundraising target: Zack Danger Brown's Kickstarter page, "Potato Salad," July 2017. www.kickstarter.com.

127 "names of idiots": Comment from YouTube user Radickly Rick. https://www.youtube.com.

129 The prize was access: Will Smale, "How Controversial Beer Firm Brewdog Became So Popular," *BBC News*, January 5, 2015.

129 "It has, among other antics": Jon Henley, "The Aggressive, Outrageous, Infuriating (and Ingenious) Rise of BrewDog," *The Guardian*, March 24, 2016.

129 Watt and Dickie combine: Ibid.

130 They now have more than forty bars: BrewDog, "International," July 2017. www.brewdog.com.

130 The first 1,329: Alanna Petroff, "Craft Beer Bet Returns 2,800% to Early Investors," *CNN Money*, April 10, 2017.

130 "The mental shackles": Bill Bruce, "BrewDog Smashes £4.25m Crowdfunding Record," *FoodBev Media*, December 23, 2013.

130 BrewDog's fourth round: Ibid.

130 As an Equity Punk: BrewDog, "#PunkAGM2016," July 2017. www.brewdog.com.

130 "completely controlled by our Equity Punk": BrewDog, "The Benefits of Investing in BrewDog," July 2017. www.brewdog.com.

131 For all their capacity: Petroff, "Craft Beer Bet."

131 "We're not going to let the deal": Ibid.

132 He also questioned how Roberts's wife: Derek Smart, "Star Citizen—Interstellar Discourse," *Derek Smart* (blog), July 10, 2015. www.dereksmart.com.

132 "He who must not": Starfarer_Rio, "'He Who Must Not Be Named' Admits to Wanting to Kill the Project," *RSI Community Forums*, October 2015. www.robertsspaceindustries.com.

132 Seventy-four percent of Citizens: Loonie, "Are Refunds for Our Packages/Ships from CIG Now Okay to Be Requested?," RSI Community Forum, July 2015. www.robertsspaceindustries.com.

132 "I'm not sure you understand": Blightbow, "Stretch Goal 48m—External Business Audit," RSI Community Forum, June 2014. www.robertsspaceindustries.com.

133 *10 for the Chairman*: *10 for the Chairman*, YouTube video, posted by "Star Citizen," www.youtube.com.

133 He wrote a long letter: Roberts, "Letter from the Chairman," July 20, 2015.

133 "Deep down most people": Alex Walker, "For Some, Star Citizen Is a Deeply Emotional Project," *Kotaku*, October 12, 2015.

133 Similarly for BrewDog you can bet: Rob Davies, "'Punk' Beer Maker Brewdog Sells 22% of Firm to Private Equity House," *The Guardian*, April 9, 2017.

134 "In July of 2013": Eliza O'Neill's GoFundMe page, "Saving Eliza," July 2017. www.gofundme.com/ElizaONeill.

134 "Hope is a nice word": "This video will save a little girl's life: #SavingEliza," YouTube video, posted by "VonWong," April 2, 2014, www.youtube.com.

134 The video was the breakthrough: Ibid.

135 Toward the end of 2015: Myriah Towner and Jill Reilly, "'She's Running Out of Time, but We Haven't Run Out of Hope,'" *Daily Mail*, November 13, 2015.

135 "Or as some would call it": Jonathan Zittrain, Twitter post, January 19, 2016, 6:32 a.m. www.twitter.com/zittrain.

136 GoFundMe has become: GoFundMe, "Top 10 Crowdfunding Sites," July 2017. www.crowdfunding.com.

136 "people understand the game": Nehkara, discussion with authors. The following Nehkara quotes also come from this source.

137 "to bring the 'verse to life": Imperial News Network, "About Us," July 2017.
www.imperialnews.network.

Chapter 8: Taking the Turn from Old to New Power

139 "Do you think that the Minister": Science and Technology Committee,
"'Boaty McBoatface' Competition Examined with NERC Representatives,"
UK Parliament, May 5, 2016.

139 "an astonishingly great outcome": Ibid.

140 "the UK's largest": Natural Environment Research Council, "About the Ship,"
July 2017. www.nameourship.nerc.ac.uk.

140 To engage the public: #NameOurShip, July 2017. www.nameourship.nerc
.ac.uk.

140 "We are excited to hear": Jonathan Amos, "Name Sought for New UK Polar
Ship," *BBC News,* March 17, 2016.

140 NERC's press release trailed: "Campaign Launched to Name the UK's State-
of-the-Art £200m Polar Research Ship," Natural Environment Research
Council, March 17, 2016.

140 "Boaty McBoatface": BBC News Jersey, "Boaty McBoatface Instigator 'Sorry'
for Ship Name Suggestion," *BBC News,* March 21, 2016.

140 "it's an absolutely brilliant name.": "$380 Million Ship May Be Named Boaty
McBoatface Thanks to an Online Poll," *Sydney Morning Herald,* March 21,
2016.

140 "I'm terribly sorry": Helena Horton, "£200m Ship Possibly to Be Named
Boaty McBoatface Thanks to an Online Poll," *Telegraph,* March 20, 2016.

141 "The people of the United Kingdom": Royal Caribbean International, "Royal
Caribbean Asks Genius Behind Boaty McBoatface to Help Name Future
Ship," *PR Newswire,* March 31, 2016.

141 The website generated 2.3 million: Bluestag, "#NameOurShip," July 2017.
www.bluestag.co.uk.

141 Of course, Boaty won: Tiffany Ap, "Landslide Win for 'Boaty McBoatface' in
$300M Research Ship Naming Poll," CNN, April 18, 2016.

141 "We want a name that lasts": Jessica Elgot, "Boaty McBoatface May Not Be
Name of New Polar Research Vessel," *The Guardian,* April 18, 2016.

141 To soften the blow: BBC, "'Boaty McBoatface' Polar Ship Named After
Attenborough," *BBC News,* May 6, 2016.

142 "very elegant compromise": Dan Bloom, "'I Voted for Boaty McBoatface':
Top Scientist Stuns MPs with Surprise Confession in Parliament," *Mirror,*
May 10, 2016.

142 "superficial, low-stakes questions": Alistair Stoddart, "Beyond #Boaty
McBoatface: Digital Tools for Democracy," *The Democratic Society,* May 20,
2016.

142 "we are only really now": Science and Technology Committee, "'Boaty McBoatface' Competition Examined with NERC Representatives."

145 "Did you always want": Emily Greenhouse, "JPMorgan's Twitter Mistake," *The New Yorker,* November 16, 2013.

146 It was suffering from sales decline: Andrew O'Connell, "Lego CEO Jørgen Vig Knudstorp on Leading Through Survival and Growth," *Harvard Business Review,* January 2009.

146 In 2003 it posted its biggest loss: David C. Robertson with Bill Breen, *Brick by Brick: How LEGO Rewrote the Rules of Innovation and Conquered the Global Toy Industry* (New York: Crown Publishing Group, 2014).

147 "I came with my family": Mary Jo Hatch and Majken Schultz, *Taking Brand Initiative: How Companies Can Align Strategy, Culture, and Identity Through Corporate Branding* (San Francisco: Jossey-Bass, 2008), 197.

147 "Take MOCpages": MOCpages, "About MOCpages," July 2017. www.moc-pages.com.

147 "visitors from around the world": Ibid.

148 "was the only adult": Robin Sather, discussion with authors, April 8, 2016. The following two quotes are from the same source.

148 "few in number and not worth listening to": Robertson, *Brick by Brick,* 249.

148 Even though AFOLs: Ibid., 251.

149 So would local media: Ibid.

149 Working with Barnes's firm: Brendan Koerner, "Geeks in Toyland," *Wired,* February 1, 2006.

149 "guy with an idea": "LEGO Architecture Designer Adam Reed Tucker—Brickworld Chicago 2014," YouTube video, 17:43, posted by "Beyond the Brick," July 28, 2014, www.youtube.com.

149 "Lego Architecture":"Architecture," July 2017. www.lego.com.

150 Lego Ambassador Network: "LEGO Ambassador Network," July 2017. www.lan.lego.com.

150 The "Lego Ideas": "Ideas," July 2017. www.ideas.lego.com.

150 Those whose Lego ideas: "LEGO Ideas—How It Works," July 2017. www.ideas.lego.com.

150 "professional female minifigures": Amelia Butterly, "Lego Makes Female Scientist Figures, After Campaign," *BBC News,* June 4, 2014.

151 The base was critical to the success: Box Office Mojo, "*The Lego Movie,*" July 2017. www.boxofficemojo.com.

151 "Is there anyone out there": Christopher Ratcliffe, "*The Lego Movie*'s Solid Social Marketing Strategy," *Econsultancy,* November 6, 2013.

151 The film's creators: Lara O'Reilly, "Lego to Co-create Marketing Campaign with UK Fans," *Marketing Week,* April 7, 2014.

151 "The Lego company was like": Robin Sather, discussion with authors.

152 As Knudstorp has noted: "Successful LEGO Strategy Delivers Continued Strong Growth," July 2017. www.lego.com.

152 "You're crashing someone else's": Yun Mi Antorini, discussion with authors, April 26, 2016.

152 "The Lego Group could never": Ibid.

153 "hard work to make it work": Ibid.

153 "World's Most Powerful Brand": "Lego Overtakes Ferrari as the World's Most Powerful Brand," *Brand Finance*, July, 2017.

154 "They say that there's two types": Ryan Lizza, "Inside the Collapse of *The New Republic*," *The New Yorker*, December 12, 2014.

156 Only four organizations in the world: Telegraph Staff, "NHS Is Fifth Biggest Employer in World," *Telegraph*, March 20, 2012.

156 "two old power systems": Helen Bevan, discussion with authors, March 22, 2016.

156 "hitting the target": Ibid.

156 "School for Health and Care Radicals": "School for Change Agents," July 2017. www.theedge.com.

157 In 2013, she and a young clinician: Damian Roland, "Has NHS Change Day Made a Difference Six Months On?," *The Guardian*, September 13, 2013.

157 In all, in its first year Change Day: Jackie Lynton and Joe McCrea, "The Difference a Day Makes . . . Interim Report for NHS Change Day—April 2013," *National Health Service*, 2013.

157 By year two, there were over 800,000: James Shrimpling, "NHS Change Day: Let's Get Back to the Grassroots," *The Guardian*, March 20, 2015.

157 "Change Day fundamentally": Helen Bevan, "Biggest Ever Day of Collective Action to Improve Healthcare That Started with a Tweet," *National Health Service*, June 14, 2013.

158 "very powerful anti-psychotic drugs": Helen Bevan, discussion with authors. The following Bevan quotes also come from this source.

158 Together, in less than three years: NHS England, "Using Social Movement to Transform the NHS—Dr. Helen Bevan," *NHS News*, January 30, 2017.

158 "community of change agents": Helen Bevan, discussion with authors.

159 "Your main responsibility is to seek": Hila Lifshitz-Assaf, "Dismantling Knowledge Boundaries at NASA: From Problem Solvers to Solution Seekers," *Administrative Science Quarterly*, forthcoming.

160 These stars recruited others: Ibid.

Chapter 9: Leadership

162 "Thank you for everything": President Obama, Twitter post, January 10, 2017, 11:52 p.m. www.twitter.com/potus44.

162 "I alone can fix it": Yoni Appelbaum, "'I Alone Can Fix It,'" *The Atlantic*, July 21, 2016.

164 "*We* are the ones": Barack Obama, "Barack Obama's February 5th Speech," *New York Times*, February 5, 2008. www.nytimes.com.

164 "ready on day one": Larry Beinhart, "Who's Ready on Day One?," *Huffington Post,* May 30, 2008.

164 "Respect. Empower. Include": Adrian Prandle, "Respect, Empower, Include— Everyday People, Extraordinary Results," *LaborList,* April 30, 2009.

164 Shepard Fairey's iconic: Laura Barton, "Hope—the Image That Is Already an American Classic," *The Guardian,* November 10, 2009.

165 Everyone had a lane: Zack Exley, "The New Organizers, What's Really Behind Obama's Ground Game," *Huffington Post,* November 28, 2008. The related quotes that follow also come from this source.

166 "Wow. Whoa. That is some": "Here's Donald Trump's Presidential Announcement Speech," *Time,* June 16, 2015.

166 He reportedly hired: Aaron Crouch and Emmett McDermott, "Donald Trump Campaign Offered Actors $50 to Cheer for Him at Presidential Announcement," *Hollywood Reporter,* June 17, 2015.

167 "'I'm f——trembling'": Abby Ohlheiser, "'We Actually Elected a Meme as President': How 4chan Celebrated Trump's Victory," *Washington Post,* November 9, 2016.

167 The 4channers appropriated: Olivia Nuzzi, "How Pepe the Frog Became a Nazi Trump Supporter and Alt-Right Symbol," *Daily Beast,* May 26, 2016.

167 In one instance, he retweeted: Taylor Wafford, "Donald Trump Retweets Racist Propaganda," *Newsweek,* November 23, 2015.

168 "If you see somebody getting ready": Philip Bump, "Donald Trump Reverses Course on Paying Legal Fees for Man Who Attacked Protester. But Could He Do It?," *Washington Post,* March 15, 2016.

168 Positive sentiment about Trump: 4C Insights, "Election Night 2016 Impact Report," November 10, 2016. www.4cinsights.com.

169 This moment, with much of the country: Natalie Andrews, "How Some Social Media Data Pointed to a Donald Trump Win," *Wall Street Journal,* November 10, 2016.

169 Following Trump's election, George Orwell's: "George Orwell Classic '1984' Gets Sales Boost After Trump Advisor Coins 'Alternative Facts,'" Associated Press, January 24, 2017.

170 "the one weird trait": Matthew MacWilliams, "The One Weird Trait That Predicts Whether You're a Trump Supporter," *Politico,* January 17, 2016.

170 "in a general election": Ibid.

170 "If you want to know how I'll govern": Tim Dickinson, "No We Can't," *Rolling Stone,* February 2, 2010.

171 "Obama must continue to feed": Exley, "The New Organizers."

171 "criminal political negligence": Dickinson, "No We Can't."

171 The administration then made: Ari Melber, "Looking Back at One Year of Organizing for America," *The Nation,* January 14, 2010.

173 "great sense of inner peace": John L. Allen, Jr., "On the Pope's 'Mystical Moment': Something Did Happen," *National Catholic Reporter,* October 6, 2013.

173   "You put it on instead": David Willey, "Pope Francis' First Moves Hint at Break with Past," *BBC News*, March 16, 2013.

173   "I'll stay down here": Mike Hammer, "A Look at Pope Francis as He Assumes His Role of the Head of the Roman Catholic Church," *Downtown Magazine*, March 14, 2013.

173   "God may bless": Vatican Radio, "Pope Francis on His Pontificate to Date," March 13, 2015.

174   He had other plans afoot: David Willey, discussion with authors, August 8, 2016.

174   The bookies didn't think: Benjamin Eckstein and Ginger Adams Otis, "Bookies Had It Wrong on Pope Selection," *NY Daily News*, March 13, 2013.

174   So he had arrived in Rome: Liz Dodd and Abigail Frymann Rouch, "Francis Hints at Short Papacy—'Maybe Just Two or Three Years,'" *The Tablet*, March 13, 2015.

174   Revelations of decades: Frank Newport, "Catholic Church Attendance Drops This Year in Midst of Scandal," *Gallup*, September 18, 2002.

174   Catholics' opinions: Pew Research Center, "Positive Impact of Pope Francis on Views of the Church, Especially Among Democrats and Liberals," *Pew Research Center*, October 7, 2015.

174   Too few young men: Center for Applied Research in the Apostate, "Frequently Requested Church Statistics," Georgetown University, July 2017.

174   Many parishes were short: Dave Breitenstein, "U.S. Catholics Face Shortage of Priests," *USA Today*, May 25, 2014.

174   In Rome, inefficiencies and inertia: John J. Allen Jr., "The Risks of Pope Francis's Never-Ending Vatican Reform," *CRUX*, December 22, 2016.

175   "terrible crisis": David Willey, discussion with authors.

175   He is even well regarded: Megan Cornwell, "Pope Francis Is World's Most Popular Leader, Poll Finds," *The Tablet*, March 28, 2016.

176   Think of the time he traded: Yasmine Haflz, "Pope Francis' Car Shows His Commitment to Humility: Catholic Leader Chooses Ford Focus," *Huffington Post*, July 15, 2013.

176   Or when he lingered: Faith Karimi, "Pope Francis' Embrace of a Severely Disfigured Man Touches World," CNN, November 7, 2013.

176   Or when he washed: Lizzie Davies, "Pope Francis Kisses Feet of Women and Muslim Man in Maundy Thursday Rite," *The Guardian*, April 17, 2014.

176   "walking parable": Father Thomas Reese, discussion with authors, August 18, 2016.

176   "inverted pyramid": Hazel Torres, "Pope Francis Calls for 'Revolution' in Catholic Church with Pope at the Bottom of Inverted Pyramid Holy See," *Church*, October 20, 2015.

177   "engaging families themselves": Pope Francis, "Address of His Holiness Pope Francis: Ceremony Commemorating the 50th Anniversary of the Institution of the Synod of Bishops," October 17, 2015.

177 "To my knowledge": Tracy Connor, "Pope Francis' Latest Surprise: A Survey on the Modern Family," NBC News, November 5, 2013.

177 Another initiative that pushed power down: Pope Francis, *Amoris Laetitia* (Rome: The Holy See, March 19, 2016).

177 "Who am I to judge?": "Pope Francis: Who Am I to Judge Gay People?," BBC News, July 29, 2013.

177 "disjointed multitude of doctrines": Rachel Zoll, "Church Could 'Fall Like a House of Cards' If It Doesn't Temper Stances on Contraception, Gays and Abortion, Pope Says," *National Post,* September 19, 2013.

178 "It's hypocrisy to call": Cindy Wooden, "Christians Who Reject All Refugees Are 'Hypocrites', Pope Says," *Catholic News Service,* October 13, 2016.

178 "Mercy is doctrine": James Carroll, "With His New Book, Pope Francis Unlocks the Door," *The New Yorker,* January 12, 2016.

179 "And trust in all four institutions": 2017 Edelman Trust Barometer, "Executive Summary," 2017, 4.

179 Thus was born: Ai-jen Poo, discussion with authors, January 10, 2017.

180 She founded the National Domestic Workers Union: Archives Division at Auburn Avenue Research Library, "Mrs. Dorothy Lee Bolden, Founder and President of the National Domestic Workers of America, Inc.," November 19, 2011. www.aarlarchives.blogspot.co; National Domestic Workers Alliance, July 2017. www.domesticworkers.org.

180 "creates power for more people": Ai-jen Poo, discussion with authors. The following Ai-jen Poo quotes and related data also come from this source.

181 That initiative has strengthened: National Domestic Workers Alliance, "Worker Organizing & Leadership," July 2017. www.domesticworkers.org.

181 so never again could you: Natt Garun, "Egg Minder Smart Tray Lets You Remotely Check the Freshness of Your Eggs," *Digital Trends,* July 5, 2013.

182 Comstock herself had championed: Steve Lohr, "Quirky, an Invention Start-Up, Files for Bankruptcy," *New York Times,* September 22, 2015.

182 Her success in leading GE: Forbes, "The World's 100 Most Powerful Women," *Forbes,* July 2017.

182 She chalked up early wins: GrabCAD, "GE Jet Engine Bracket Challenge," *GrabCAD Community,* July 2017. www.grabcad.com.

183 The winner was a young: Ibid.

183 Within GE, Comstock called: Eric Ries, *The Lean Startup: How Today's Entrepreneurs Use Continuous Innovation to Create Radically Successful Business* (New York: Crown Publishing Group, 2011).

183 This led to the creation: GE, "GE//FastWorks," *Innovation Benchmark,* March 11, 2016.

183 A project to create a digital wind farm: GE, "GE 2015 Integrated Summary Report," 2015, p. 19.

183 GE has estimated that FastWorks: "Bridging Worlds: The Future Role of the Healthcare Strategist," Society for Healthcare Strategy & Marketing Development of the American Hospital Association, 2014, 20.

183 "distribute the decision-making process": Beth Comstock, "Welcome to the Emergent Era," *Medium*, December 1, 2016.

184 "While she has a very big job": Peter Sims, discussion with authors, week of January 15, 2017.

184 The result, LittleMonsters.com: Little Monsters, July 2017. www.little monsters.com.

184 "Our bond is beyond": Billboard Staff, "Lady Gaga Fan on Being a Little Monster: 'We Are Genuinely Trying to Make the World a Better Place,'" *Billboard*, August 18, 2015.

184 "doesn't judge or criticize": Melissa A. Click, Hyunji Lee, and Holly Willson Holladay, "Making Monsters: Lady Gaga, Fan Identification, and Social Media," *Popular Music and Society* 36 (2013).

185 "Rather than Gaga pushing": Amber L. Davisson, *Lady Gaga and the Remaking of Celebrity Culture* (London: McFarland & Co., 2013), 50.

185 In 2010, Gaga: Elisa Vliebeck, "Reid to Lady Gaga: 'Don't Ask, Don't Tell' Vote Will Be Next Week," *The Hill*, September 14, 2010.

185 With Gaga's call to action: Nancy Scola, "With Call to Action, Obama Brings Down the House," *The Atlantic*, July 26, 2011.

185 "U.S. Senate Getting Gaga-Compliant": Nancy Scola, "U.S. Senate Getting Gaga-Compliant Phone System (and Visual Voicemail!)," *TechPresident*, September 21, 2010.

186 "female leaders, on the average": Alice H. Eagly, "Women as Leaders: Leadership Style Versus Leaders' Values and Attitudes," *Harvard Business School Publishing*, 2013, 4.

186 "power-over": Mary Parker Follett, *The New State: Group Organization the Solution of Popular Government* (University Park: Penn State University Press, 1998), xviii.

186 This idea presages: Amy Allen, "Feminist Perspectives on Power," *Stanford Encyclopedia of Philosophy*, October 19, 2005. www.plato.stanford.edu.

186 "It began with one voice": Starbucks, "What 'Race Together' Means for Starbucks Partners and Customers," Starbucks Newsroom, March 16, 2015.

187 "The last few weeks I've felt": Ibid.

187 "Being a barista": Scott Kleinberg, "Starbucks #RaceTogether Campaign Brews Up Bitter Social Media Reaction," *Chicago Tribune*, March 18, 2015.

187 "Not sure what @Starbucks": Rebecca Cullers, "The Internet Is United in Despising Starbucks' 'Race Together' Cup Campaign," *AdWeek*, March 18, 2015.

187 "as planned": Agence France-Presse, "Starbucks Halts Race Campaign amid 'Cascade of Negativity,'" Inquirer.net, March 24, 2015.

187 "We made a tactical mistake": Austin Carr, "The Inside Story of Starbucks's Campaign, No Foam," *Fast Company*, June 15, 2015.

188 He then partnered: Howard Schultz and Larry Kramer, "Why Race Together? Because Diversity Matters," *USA Today*, March 17, 2015.

189 "3 black women": Black Lives Matter, Twitter post, February 5, 2015, 8:26 a.m. www.twitter.com/blklivesmatter.

189 Garza's friend Patrisse Cullors: Julia Craven, "Black Lives Matter Co-founder Reflects on the Origins of the Movement," *Huffington Post*, September 30, 2015.

189 It soon began to catch on: Alicia Garza, "A Herstory of the #BlackLivesMatter Movement by Alicia Garza," *Feminist Wire*, October 7, 2014. Subsequent quotes in this paragraph are from the same source.

190 The three women: Akiba Solomon, "Get on the Bus: Inside the Black Life Matters 'Freedom Ride' to Ferguson," *Colorlines*, September 5, 2014.

190 This 50,000-strong march: Olivia Fleming, "#BlackLivesMatter: These Two Young Women Rallied 50,000 to Protest Police Brutality in NYC," *Elle*, December 18, 2014.

191 "Prospective chapters must submit": Jelani Cobb, "The Matter of Black Lives," *The New Yorker*, March 14, 2016.

191 "The other piece of this involvement": Monica J. Casper, "Black Lives Matter / Black Life Matters: A Conversation with Patrisse Cullors and Darnell L. Moore," *truthout*, December 3, 2014.

191 "Black Lives Matter affirms the lives": Black Lives Matter, "About the Black Lives Matter Network," July 2017. www.blacklivesmatter.com.

191 Jackson found himself booed: Cobb, "The Matter of Black Lives."

192 "If it is a protest": Alicia Lu, "Did Al Sharpton 'Monopolize' Black Lives Matter Protests for His Own Gain?," *Bustle*, December 21, 2014.

192 "the phenomenon around DeRay": Alicia Garza, in discussion with the authors, October 30, 2017.

Chapter 10: The Art of Blending Power

197 Just two months after the recall election: Lyn Bartels and Kurtis Lee, "Evie Hudak Resigns: Colorado State Senator Avoids Recall Election," *Denver Post*, November 27, 2013.

197 Even when public opinion: Carl Bialik, "Most Americans Agree with Obama That More Gun Buyers Should Get Background Checks," *FiveThirtyEight* (blog), January 5, 2016. www.fivethirtyeight.com.

198 After all, the NRA: BBC, "US Gun Control: What Is the NRA and Why Is It So Powerful?," *BBC News*, January 8, 2016.

198 The two state senators: Nancy Watzman and Anu Narayanswamy, "Did Guns Beat Money in Colorado Recalls?," *Sunlight Foundation*, December 10, 2013.

198 "The reason my brother and I": Eli Stokols, "Historic Colorado Recalls Engineered by Political Newcomers," Fox 31 Denver, September 11, 2013.

199 "The benefit of pushing": Mark Glaze, discussion with authors, April 13, 2017. The following Glaze quotes are from the same source.

200 But to beat the Manchin-Toomey bill: Molly Moorehead, "A Summary of the Manchin-Toomey Gun Proposal," *PolitiFact,* April 30, 2013.

200 Before this bill: Robert Draper, "Inside the Power of the N.R.A.," *New York Times,* December 12, 2013.

200 "Tellingly, the NRA had declined": Douglas E. Schoen, "The Myth of NRA Power: What the NRA Really Means in American Politics Today," *Medium,* January 6, 2016.

201 *return on fear:* Ibid.

201 "I think I always": Manu Raju, "Heitkamp Defends Gun Vote," *Politico,* April 24, 2013.

201 Polls show that more than 90 percent: Tim Dickinson, "Gun Control: 45 Percent of the Senate Foils 90 Percent of America," *Rolling Stone,* April 17, 2013.

202 "Marksmanship Programs Your Child Will Love": Emily Rupertus, "Marksmanship Programs Your Child Will Love," *NRA Blog,* August 16, 2016. www .nrablog.com.

202 "The gun control side": Jennifer Carlson, discussion with authors, August 15, 2016.

202 "By stoking nationalist claims": Mark Glaze, discussion with authors.

203 "The only way we stop this": Zack Beauchamp, "This Chilling NRA Ad Calls on Its Members to Save America by Fighting Liberals," *Vox,* June 29, 2017.

203 It's instructive that: Everytown for Gun Safety, July 2017. www.everytown .org.

203 "Moms Demand Action for Gun Sense in America": Moms Demand Action for Gun Sense in America, July 2017. www.momsdemandaction.org.

203 And in recent years, a more grassroots: Everytown, "Everytown Claims Major Victory in Washington State with Passage of I-594, Proving That When Citizens Take Public Safety to the Ballot Box, Gun Safety Wins," November 4, 2014. www.everytown.org.

203 62,000 calls to the Senate: Everytown, email to the authors, July 10, 2017.

204 Every second of every day: TED Staff, "TED Reaches Its Billionth Video View!," *TED Blog,* November 13, 2012. www.blog.ted.com.

204 "a growing and influential audience": TED, "History of TED," July 2017. www.ted.com.

205 Attendance at today's flagship TED event: TED, "TED2018," July 2017. www .ted2018.ted.com.

205 Anderson took over: Mick Brown, "'I Was Losing $1 Million a Day, Every Day for 18 Months': Meet Chris Anderson, the Man Behind TED Talks," *The Telegraph,* April 26, 2016.

205 The first talks were posted: TED, "History of TED."

206 "to illuminate, clarify, engage": Chris Anderson, "TED Isn't a Recipe for 'Civilisational Disaster,'" *The Guardian,* January 8, 2014.

207 Thousands of events have taken place: TED, "Organize a Local TEDx Event," July 2017. www.ted.com.

207 "It didn't matter if you were": Lara Stein, discussion with authors, September 9, 2016.

207 In the official "rules": TED, "TEDx Rules," July 2017. www.ted.com.

208 "vortex-based mathematics": Nilofer Merchant, "When TED Lost Control of Its Crowd," *Harvard Business Review,* April 2013.

208 "a mockery . . . that hurt": Ibid. Ellipses in original.

208 "The reason behind that": Randy Bretz, discussion with authors, July 22, 2016.

209 But this fourth TED community: TED, "Stories for 'Open Translation Project,'" www.ted.com.

209 The project has now seen: TED, "TED Translators," July 2017. www.ted.com.

209 He has racked up: Jenny Zurawell, "Meet Sebastian Betti, TED Volunteer Translator," *TED Blog,* www.blog.ted.com.

211 "We do our best thinking":"2016 Engineering Career Fair Collaborative—Participants: Local Motors," Harvard University, November 16, 2016. www.harvard-csm.simplicity.com.

211 "so radical that it's hard": Kristine Owram, "How Local Motors Hopes to Disrupt the Auto Industry with a $53,000 3D-Printed Car," *Financial Post,* January 19, 2016.

211 "Any customer can have a car": Henry Ford, *My Life and Work: An Autobiography of Henry Ford* (NuVision Publications, 2009).

211 "We asked you, the Local Motors Community": Will Schmidt, "This Is How You Change and Disrupt the Paradigm of Auto Manufacturing," *Tech.co,* June 8, 2015.

212 "I actually kind of styled it": Patty Hastings, "Vancouver Man Designs First 3-D Printed Car," *The Columbian,* August 13, 2015.

212 "We're looking for those": Alex Fiechter, discussion with authors, August 6, 2016.

213 "drawing the box tightly": John B. (Jay) Rogers, discussion with authors, July 23, 2016.

213 "I wanted to validate this": Nicolas de Peyer, "Entry: Reload Redacted—Swim and Sport," Local Motors Labs, July 2017. www.launchforth.io.

213 "There wasn't a single critique": Jay Rogers, discussion with authors.

213 "Sometimes we've got people": Ibid.

215 Just four months later: European Voice, "Spain: European Parliament results," *Politico,* May 23, 2014.

215 In the Spanish general election: Luke Stobart, "A Year of Change Postponed?," *Jacobin,* December 18, 2015.

215 Its philosophy—as a party leader: Jorge Moruno and Carlos Declos, "'Our Situation Is Quixotic and Machiavellian': An Interview with Podemos' Jorge Moruno," *Open Democracy,* December 16, 2015.

215 Youth unemployment: Gary Younge, "Spain's Unemployed: One in Five Under 30 Still Looking for That First Job," *The Guardian,* March 30, 2011.

215   "City square by city square": Diego Beas, "How Spain's 15-M Movement Is
      Redefining Politics," *The Guardian*, October 15, 2011.

215   In the November national elections: Alistair Dawber, "Pablo Iglesias: How
      the Leader of the Leftist Podemos Party Upset Spain's Elites to Reach the
      Brink of Power," *The Independent*, December 25, 2015.

216   "sneering response of the old elites": Pablo Iglesias, "Understanding
      Podemos," *New Left Review* 93 (May–June 2013).

216   A lifelong activist: Raphael Minder, "Spanish Upstart Party Said It Could,
      and Did. Now the Hard Part Begins," *New York Times*, May 29, 2014.

216   The show was called *La Tuerka*: *La Tuerka*, YouTube videos, posted by
      "LaTuerka," www.youtube.com.

216   "TV studios have become": Iglesias, "Understanding Podemos." Subsequent
      Iglesias quotes, until the 2014 speech, are from this source.

217   Soon, across the country: Matt Moffett, "Spanish Voters Weary of Scandal-
      Ridden Political Class Known as 'La Casta,'" *Wall Street Journal*, Decem-
      ber 18, 2015.

218   "horizontality and transparency": "Podemos—Translated Manifesto," *Cun-
      ning Hired Knaves* (blog), January 20, 2014. www.hiredknaves.wordpress
      .com.

218   The number was reached: Bécquer Seguín and Sebastiaan Faber, "Can
      Podemos Win in Spain?," *The Nation*, January 14, 2015.

218   Circles began to spark: Giles Tremlett, "The Podemos Revolution: How a
      Small Group of Radical Academics Changed European Politics," *The Guard-
      ian*, March 31, 2015.

218   "hacker logic": Andrew Dolan, "Podemos: Politics by the People," *Red Pep-
      per* (blog), February 22, 2015. www.redpepper.org.

218   "To create a branch": Ibid.

219   Five candidates: Dan Hancox, "Can They?," *London Review of Books* 37,
      no. 24 (December 17, 2015): 25–26.

219   "popular participation in the most important decisions": Sonia Bussu, "A
      New Age of Party Politics?," *Involve*, June 10, 2015.

219   "unprecedented exercise of direct democracy": Seguín and Faber, "Can
      Podemos Win in Spain?"

219   An online debating site: Ibid.

219   By December 2015, Podemos: Sara Prim, "The PP Wins the Spanish Elec-
      tions but Loses Its Absolute Majority," *Catalan News*, December 21, 2015.

220   It had broken the country's: Alistair Dawber, "Spain Elections: Madrid
      Goes Wild for Podemos Leader Pablo Iglesias but Has the Party Started Too
      Soon?," *The Independent*, December 21, 2015.

220   "We Are All Khaled Said": We Are All Khaled Said's Facebook page, July
      2017. www.facebook.com.

220   "I once said": Wael Ghonim, "Let's Design Social Media That Drives Real
      Change," TED video, 13:34, December 2015. www.ted.com.

220 "The euphoria faded": Ibid.

221 The party has seen successes: Oscar Reyes, "Spain's Democratic Spring: How the Movements Stood for Mayor—and Won," *Red Pepper,* July 1, 2015. www .redpepper.com.

Chapter 11: New Power at Work

222 "Oh, you know, I met Bill": Reuven Gorsht, discussion with authors, January 26, 2016.

225 Research from the Young Entrepreneur Council: Rob Ashgar, "Study: Millennials Are the True Entrepreneur Generation," *Forbes,* November 11, 2014.

225 "more than half (53%)": "Millennial Impact Report: Inspiring the Next Generation Workforce," *Achieve,* 2014.

226 "trophy case": Lorenzo Ligato, "Here's How to Unlock All of the New Snapchat Trophies," *Huffington Post,* October 20, 2015.

226 The online service TINYpulse: TINYpulse, July 2017. www.tinypulse.com.

226 It has been adopted: Ibid.

227 "The Founders Generation": David Sims, "All Hail 'The Founders,'" *The Atlantic,* December 2, 2015.

228 Our story hits its climax: Jeff Bercovici, "Facebook Tried to Buy Snapchat for $3B in Cash. Here's Why," *Forbes,* November 13, 2013.

228 The company goes public: Portia Crowe, "Snap Is Going Public at a $24 Billion Valuation," *Business Insider,* March 1, 2017.

228 "He just wants": Austin Carr, "What Snapchat's High-Profile Exec Departures Really Tell Us About CEO Evan Spiegel," *Fast Company,* October 20, 2015.

228 A Bentley University survey: "Millennials at Work," Bentley University, November 11, 2014, 13.

230 It is the era of the "side hustle": CareerBuilder, "Millennials Significantly Outpacing Other Age Groups for Taking on Side Gigs," *PR Newswire,* September 29, 2016.

230 The reality that three: Deborah Gage, "The Venture Capital Secret: 3 Out of 4 Start-Ups Fail," *Wall Street Journal,* September 20, 2012.

230 A survey by ManpowerGroup: "Millennial Careers: 2020 Vision Facts, Figures and Practical Advice from Workforce Experts," ManpowerGroup, 2016.

231 But Buurtzorg: Buurtzorg Nederland, "Our Story," July 2017. www .buurtzorgusa.org.

232 Ten years on, Buurtzorg has: Tamsin Fulton, "Exploring the Buurtzorg Model of Care," *Public World,* December 8, 2016.

232 "a 50 percent reduction": Nina Schnider, "Reinventing Organizations: Buurtzorg and Ashoka Fellow Jos De Blok," *Ashoka,* March 7, 2017. www .ashoka.org.

232 And Buurtzorg has been repeatedly: "Home Care by Self-Governing Nursing Teams: The Netherlands' Buurtzorg Model," The Commonwealth Fund, 2015.

232 Research by the International Council of Nurses: Linda H. Aiken, Douglas M. Sloane, Luk Bruyneel, Koen Van den Heede, and Walter Sermeus, "Nurses' Reports of Working Conditions and Hospital Quality of Care in 12 Countries in Europe." http://onlinelibrary.wiley.com/doi/10.1111/inr.12211/full.

232 This culture is costly: Press Association, "Undervalued and Overworked: Nurses Across the UK Placed in 'Intolerable' Situations, New Poll Reveals," *Wales Online,* October 5, 2015.

233 "I had this meeting": Madelon von Tilburg, discussion with authors, November 20, 2016.

233 "When you start a new team": Ibid.

233 "Sometimes you find out about": Ibid.

234 In place of six layers: Alieke van Dijken and Tamsin Fulton, "The Buurtzorg Way: Happy Clients and Happy Nurses," *Happy Conference,* September 2, 2016.

234 "How do you manage": Paul J. Zak, *Trust Factor: The Science of Creating High-Performance Companies* (New York: AMACOM, 2017), 92.

234 BuurtzorgWeb: "Buurtzorg Web," July 2017. www.buurtzorg.com.

235 "new way of running": Benjamin Snyder, "Holacracy and 3 of the Most Unusual Management Practices Around," *Fortune,* June 2, 2015.

235 "As a Partner assigned": Holacracy, "Holacracy Constitution," Holacracy-One, LLC, 2013. www.holacracy.org.

236 "getting in the way of the work": Aimee Groth, "Zappos Is Struggling with Holacracy Because Humans Aren't Designed to Operate like Software," *Quartz,* December 21, 2016.

236 "It felt like being part of a code": Julia Culen, "Holacracy: Not Safe Enough to Try," *Medium,* June 27, 2015.

236 "During the next year": Deloitte, "The Future of the Workforce: Critical Drivers and Challenges," Deloitte, July 2016, 6. https://www2.deloitte.com.

237 "What's the next job": Richard Feloni, "Why LinkedIn's Head of Recruiting Asks Every Job Candidate What They Want to Do After LinkedIn," *Business Insider,* January 13, 2017.

237 Instead we should frame: Reid Hoffman, Ben Casnocha, and Chris Yeh, *The Alliance: Managing Talent in the Networked Age* (Boston: Harvard Business Review Press, 2014).

238 Here are some of the clauses: Reid Hoffman, Ben Casnocha, and Chris Yeh, "Statement of the Alliance," *The Alliance Framework,* 2014. www.thealliance framework.com.

239 In a world of fraying loyalties: Elaine Pofeldt, "Shocker: 40% of Workers Now Have 'Contingent' Jobs, Says U.S. Government," *Forces,* May 25, 2015.

240 "You eat a coffee for lunch": Ellen Scott, "People Are Not Pleased with Fiverr's Deeply Depressing Advert," *Metro*, March 10, 2017.

240 "The campaign positions Fiverr": DCX Growth Accelerator, "Fiverr Debuts First-Ever Brand Campaign," *PR Newswire*, January 9, 2017.

240 "I'd guess that plenty": Jia Tolentino, "The Gig Economy Celebrates Working Yourself to Death," *The New Yorker*, March 22, 2017.

240 With more than 24 million members: "Company Overview," July 2017. www .care.com.

240 "If we want to sustain": Sheila Marcelo, discussion with authors, September 20, 2016.

241 "Fair Care Pledge": "Take the Fair Care Pledge Today!," July 2017. www.fair carepledge.com.

241 Now they are advocating on Capitol Hill: Julia Quinn-Szcesuil, "What Caregivers Need to Know About the Domestic Workers' Bill of Rights," *Care.com*, July, 2017. www.care.com.

Chapter 12: The Future: A Full-Stack Society

242 "Carr-Benkler wager": Rick Kazman, "The Carr-Benkler Wager and Its Implications for ULS Software Engineering," Association for Computing Machinery, May 10, 2008.

242 "nonmarket-based cooperation": Yochai Benkler, *The Wealth of Networks* (New Haven, CT: Yale University Press, 2006).

242 "techno-anarcho-utopian magnum opus": Matthew Ingram, "The Carr-Benkler Wager and the Peer-Powered Economy," *Gigaom*, May 9, 2012.

243 "amateur activity springing up": Olivier Silvian, "Contingency and the 'Networked Information Economy': A Critique of *The Wealth of Networks*," *International Journal of Technology, Knowledge, and Society* 4 (2008): 7.

243 "We could decide to appoint": Yochai Benkler, "Carr-Benkler Wager Revisited," *Yochai Benkler's blog*, May 7, 2012. www.blogs.harvard.edu.

243 "Pay Up, Yochai Benkler!": Nicholas Carr, "Pay Up, Yochai Benkler," *Rough Type* (blog), May 1, 2012. www.roughtype.com.

243 "For investors, no less than": Benkler, "Carr-Benkler Wager Revisited."

244 The early flourish of something like: Couchsurfing International, July 2017. www.couchsurfing.com.

244 "I'm thrilled that you're writing": Yochai Benkler, discussion with authors, December 2, 2016.

244 "What's interesting is that if you teach": Ibid.

246 "Benkler's dream": Benkler, "Carr-Benkler Wager Revisited."

246 "Kickstarter is not a store": Strickler, Chen, and Adler, "Kickstarter Is Not a Store."

246 "public benefit corporation": Yancey Strickler, Perry Chen, and Charles Adler, "Kickstarter Is Now a Benefit Corporation," *Kickstarter* (blog), September 21, 2015. www.kickstarter.com.

246 Kickstarter's charter boldly: "Charter," July 2017. www.kickstarter.com.

247 "We told people": Perry Chen, discussion with authors, March 10, 2017.

247 A study at the University of Pennsylvania: University of Pennsylvania Wharton School of Business, "Wharton Crowdfunding Study," July 2017. www .crowdfunding.wharton.upenn.edu.

247 The platform has generated: Yancey Strickler, "Kickstarter's Impact on the Creative Economy," *Kickstarter* (blog), June 28, 2016. www.kickstarter.com.

247 "Really it's to allow": Perry Chen, discussion with authors.

247 "platform co-ops": Nathan Schneider, discussion with authors, March 6, 2017.

248 "Wall Street's economy": Ibid.

248 "A community-owned Twitter": "Notice of Annual Meeting of Stockholders," May 22, 2017. www.twitter.com.

249 "Think more artist respect": Stocksy United, "Raising the Bar—and the Industry's Expectations—of Stock Photography and Cinematography," July 2017. www.stocksy.com.

250 "online social networking will be more immune": Ching-man Au Yeung, Ilaria Liccardi, Kanghao Lu, Oshani Seneviratne, and Tim Berners-Lee, "Decentralization: The Future of Online Social Networking," In *W3C Workshop on the Future of Social Networking Position Papers*, 2009.

250 Berners-Lee's Solid project: "What Is Solid?" July 2017. https://solid.mit.edu.

251 You might know the Blockchain: BlockGeeks, "What Is Blockchain Technology? A Step-by-Step Guide for Beginners," *BlockGeeks,* July 2017. www .blockgeeks.com.

251 "It offers a way for people": The Economist Staff, "The Great Chain of Being Sure About Things," *The Economist,* October 31, 2015.

252 In a *Guardian* poll in 2017: Olivia Solon: "Americans 'Evenly Split' Over Need to Regulate Facebook and Other Big Tech," *The Guardian,* November 1, 2017.

252 "People were promised that the currents": Geoff Mulgan, "Thesis, Antithesis and Synthesis: A Constructive Direction for Politics and Policy After Brexit and Trump," *Nesta,* February 17, 2017.

252 The "visceral" difference: British Election Study team, "Brexit Britain: British Election Study Insights from the Post-EU Referendum Wave of the BES Internet Panel," *British Election Study,* June 10, 2016.

253 "deaths of despair": Shawn Donnan, "'Deaths of Despair' Surge Among US White Working Class," *Financial Times,* March 23, 2017.

253 Even the so-called "winners": Edelman, "2017 Edelman Trust Barometer Annual Global Study."

255 It is salacious and exciting: Rob Wijnberg, discussion with authors, April 15, 2016.

255 This new publication would reinvest: Ernst-Jan Pfauth, "How We Turned a World Record in Journalism Crowd-Funding into an Actual Publication," *Medium,* November 27, 2013.

255 His creation, *De Correspondent*: Ernst-Jan Pfauth, "*De Correspondent* Now Has 50,000 Paying Members," *Medium*, January 23, 2017.

256 "The people previously known": Ernst-Jan Pfauth, "Why We See Journalists As Conversation Leaders and Readers as Expert Contributors," *Medium*, April 30, 2014.

256 Readers are given tools: Ernst-Jan Pfauth, "From Councillors to Porn Actresses: Why Readers Can Soon Edit Our Site," *Medium*, January 12, 2015.

256 "steers the contributions": Pfauth, "Why We See Journalists."

256 "the world's greatest rolodex": Jelmer Mommers, "*De Correspondent*: A New Kind of Journalism," *The Coral Project*, May 3, 2017.

256 "the Netherlands' largest group interview": Dick Wittenberg and Greta Riemersma, "Seven Things the Dutch Need to Understand About How Refugees Here Feel," *De Correspondent*, December 6, 2016.

257 "*Dear Shell employees*": Jelmer Mommers, "Dear Shell Employees: Let's Talk," *De Correspondent*, February 16, 2016.

257 "The updates Jelmer published": Ernst-Jan Pfauth, "How Reader Engagement Helped Unearth the Shell Tape," *Medium*, March 6, 2017.

257 It has retention rates: Ernst-Jan Pfauth, "Dutch Journalism Platform *The Correspondent* Reaches Milestone of 40,000 Paying Members," *Medium*, December 1, 2015.

258 "The message was essentially": Claire Richard, "Audrey Tang, Brilliant Programmer, 'Hacks' Politics in Taiwan," *Open Culture*, October 12, 2015.

258 They called it g0v.tw: g0v.tw, July 2017. www.g0v.tw.

258 "If we had asked people to copy": Richard, "Audrey Tang, Brilliant Programmer."

258 "It takes five seconds": Ibid.

258 The project attracted 9,000: Ibid.

258 A new law emerged: Max Rashbrooke, "How Taiwan Is Inoculating Itself Against the Uber 'Virus,'" *CityMetric*, February 8, 2017.

259 In her journey: Ibid.

259 "unbundling democracy": Nathan Schneider, discussion with authors.

259 Participatory budgeting is an idea: Deepti Bhatnagar, Animesh Rathore, Magüi Moreno Torres, and Parameeta Kanungo, "Participatory Budgeting in Brazil," *World Bank Empowerment Case Studies*, July 2017.

260 Paris's ambitious participatory budgeting: Victoria Boelman, "Félicitations Madame Mayor: Participatory Budgeting in Paris Hits New Highs," *Nesta* (blog), October 17, 2016.

260 This mirrors the success: Wietse Van Ransbeeck, "What Is Citizensourcing?," *CitizenLab*, October 20, 2015.

260 "Blockbuster drug": Leonard Kish, "The Blockbuster Drug of the Century: An Engaged Patient," *Health Standards*, August 28, 2012.

260 Take the volunteer emergency: GoodSAM, July 2017. www.goodsamapp.org.

262 Building the full-stack society: Larry Diamond, "It Could Happen Here," *The Atlantic,* October 19, 2016.

262 The dominant early contributions: Anthony Cuthbertson, "Reddit Place: The Internet's Best Experiment Yet," *Newsweek,* April 11, 2017.

262 "what was really amazing was": Ibid.

263 More than a million people laid out: Ibid.

INDEX

(Page references in *italics* refer to illustrations.)